Studies in Modern Chemistry

Advanced courses in chemistry are changing
rapidly in both structure and content. The changes
have led to a demand for up-to-date books that
present recent developments clearly and concisely.
This series is meant to provide advanced students
with books that will bridge the gap between the
standard textbook and the research paper. The
books should also be useful to a chemist who
requires a survey of current work outside his own
field of research. Mathematical treatment has been
kept as simple as is consistent with a clear
understanding of the subject.
Careful selection of authors actively engaged in
research in each field, together with the guidance of
four experienced series editors, has ensured that
each book satisfies the needs of persons seeking a
comprehensible and modern treatment of rapidly
developing areas of chemistry.

William C. Agosta, The Rockefeller University
R. S. Nyholm, FRS, University College London

Consulting Editors

Academic editor for this volume

Lord Tedder, FRSE, University of St. Andrews

Studies in Modern Chemistry

Carbenes
nitrenes
and arynes

T. L. Gilchrist
University of Leicester
C. W. Rees
University of Leicester

APPLETON-CENTURY-CROFTS
EDUCATIONAL DIVISION
New York MEREDITH CORPORATION

Appleton-Century-Crofts
Educational Division
Meredith Corporation
440 Park Avenue South
New York, N.Y. 10016

Printed in Great Britain

Contents

Preface

Carbenes, nitrenes, and arynes are of great interest to chemists because of the central role they play, as the key intermediates, in a wide range of chemical reactions. Their study has illuminated many aspects of synthetic and mechanistic chemistry. Their organic chemistry has developed enormously over the last ten to fifteen years so that now a fairly clear pattern of reactions and reactivity, strikingly similar for all three, has emerged. Partly because of this, but also because of the many unsolved problems and gaps in our present knowledge, it now seems appropriate to present a comparative account of their chemistry.

Our approach is a simple, descriptive one that nevertheless attempts to bring the reader as nearly up-to-date as possible; there is considerable emphasis on recent work in the text and the problems. We have been selective rather than comprehensive, have concentrated on underlying principles, and have tried to emphasize what can, and what cannot, reasonably be claimed on the basis of the available experimental evidence. (The development of this area of organic chemistry could well serve as a cautionary tale in this respect.)

We have written the book primarily at the advanced student level but it should also be useful to postgraduates students and to more experienced chemists. Some of the problems are at a rather more advanced level, the aim being to extend the reader's knowledge and understanding and to direct him to some recent original literature. Solutions to the problems, with which the reader will not always agree, are usually given in the references cited.

Carbenes, benzynes, and nitrenes have already been quite extensively, but separately, reviewed, and we list the more important references on page 128. In general we have not repeated references to the original literature that are readily available from these reviews, but have given references to later work at the end of each chapter.

T. L. Gilchrist
C. W. Rees

1 Structure and relationship of the intermediates

1–1 Definitions and nomenclature

Carbenes. These are neutral, bivalent carbon intermediates in which a carbon atom has two covalent bonds to other groups and two non-bonding orbitals containing two electrons between them. If the two electrons are spin-paired, then the carbene is a *singlet*. This follows from the fact that the spin multiplicity is given by $2S + 1$, where S is the total spin, and for two spin-paired electrons, $S = 0$. If the spins of the electrons are parallel, then the carbene is a *triplet* ($S = \frac{1}{2} + \frac{1}{2} = 1; 2S + 1 = 3$).

The possible arrangements of two electrons between two orbitals of different energy are as shown:

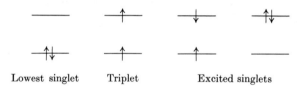

Lowest singlet   Triplet   Excited singlets

A reasonable structure for a carbene is a bent sp^2 hybrid [1]. A carbene in the lowest singlet state, with a structure of this sort, resembles a carbonium ion [2]; a triplet carbene with this structure resembles a

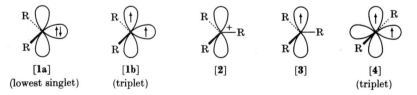

| [1a] | [1b] | [2] | [3] | [4] |
| (lowest singlet) | (triplet) | | | (triplet) |

free radical [3]. An alternative structure for the triplet carbene, and for the excited singlet, is a linear sp hybrid [4]. Structures intermediate between [1] and [4] are also possible for carbenes.

The simplest carbene, $:CH_2$, is usually called methylene. Derivatives are sometimes called methylenes; thus, $:CHCl$ is chloromethylene, and so on. More commonly, the derivatives are named chlorocarbene, phenylcarbene, and so on. Cyclic carbenes are most conveniently named by using the suffix -ylidene; for example,

H₂C
 \
  | C: is cyclopropylidene.
 /
H₂C

Nitrenes. These are the nitrogen analogues of carbenes; they are neutral, univalent nitrogen intermediates. Singlet and triplet structures are possible for nitrenes just as for carbenes, with one of the covalent bonds of the carbene replaced by the nitrogen lone pair of electrons.

The 'nitrene' nomenclature parallels that for carbenes, and is now the most widely used. In this system, : N̈H is nitrene, : N̈Ph is phenylnitrene, and so on. Other names used for these intermediates include 'imidogens', 'azenes', 'azacarbenes', and 'imenes'.

Arynes. These are neutral intermediates derived from an aromatic system by the removal of two *ortho* substituents, leaving two orbitals with two electrons distributed between them. The analogy between arynes and carbenes is not so immediately apparent, but the intermediates are structurally related because both have two electrons distributed between two available orbitals. Thus, singlet and triplet structures can be written for arynes, as for carbenes. Possible structures for benzyne are shown. Since the two orbitals are on adjacent carbons, there is likely to be some overlap between them, and we

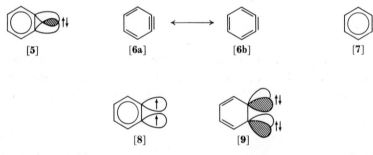

might expect the lowest singlet structure [5] to best represent benzyne. Structures [6a] and [6b] are valence-bond equivalents of [5]; they indicate the increased bond order of the aryne bond. [7] is a combination of these. Structures [8] and [9] represent possible excited states of benzyne. In the bis-carbene [9] the aromaticity of the benzene ring is lost, so the state is probably of very high energy.

The name 'benzyne' is derived from formula [6a]; related species are called arynes by analogy. The nomenclature is open to the criticism that it implies an acetylenic bond in the molecule, which there is not, although the C–C bond is shortened. An alternative is to use the prefix 'dehydro': dehydrobenzene, 1,2-dehydronaphthalene, and so on. This is strictly more accurate and is often more convenient in complex systems.

1–2 Substituents and structure

A big factor in determining the chemical behaviour of carbenes, nitrenes, and arynes is the relative energies of the different possible structures. As we have shown, it is reasonable to expect that benzyne and similar arynes will have a singlet ground state, but there is no simple and obvious way of predicting the ground states of carbenes and nitrenes. For example, the ground state of methylene probably approximates to a structure such as [**4**; R = H], whereas that of di-fluorocarbene is best represented by the singlet structure [**1a**; R = F]. There are two approaches to the problem: direct detection of the ground state and excited states by spectroscopic means, and molecular-orbital calculations. The latter may only indicate the preferred geometry for the intermediates rather than the spin states. Electron spin resonance spectroscopy has been particularly useful in detecting the triplet ground states of several carbenes and nitrenes. The inter-mediates are generated photochemically in a frozen glass at very low temperatures. Optical spectroscopy and microwave spectroscopy have also been used. Hoffmann and his co-workers have done mole-cular-orbital calculations on several carbenes.[1] The results of the experimental observations and of the calculations indicate that most carbenes have a non-linear triplet ground state [**1b**]. Exceptions include dihalogenocarbenes and carbenes with an oxygen, nitrogen, or sulphur attached to the bivalent carbon, for which the ground state is probably a singlet.* We might rationalize this in valence-bond terms by indicating possible stabilization of the singlet through dipolar resonance structures:

$$\ddot{X}\!\!\diagdown\!\!\diagup C\!: \quad\longleftrightarrow\quad {}^{+}\!X\!\!\diagdown\!\!\diagup \bar{C}\!:$$

Carbon monoxide and isocyanides are molecules that, though formally carbenes, are much more accurately represented by the dipolar resonance structures. They are nucleophilic, and the vibra-tional stretching frequencies indicate that the bonds between the carbon and oxygen or nitrogen are close to triple bonds:

$$\ddot{O}\!=\!C\!: \quad\longleftrightarrow\quad {}^{+}\!O\!\equiv\!\bar{C}\!:$$

$$R\!-\!\ddot{N}\!=\!C\!: \quad\longleftrightarrow\quad R\!-\!{}^{+}\!N\!\equiv\!\bar{C}\!:$$

* See also the footnote to Section 5–3.

To a lesser extent, the same applies to aminocarbenes and alkoxy-carbenes, which can act as nucleophiles as well as electrophiles:

$$\begin{array}{c} RO \\ \diagdown \\ R' \diagup \end{array} C: \quad \longleftrightarrow \quad \begin{array}{c} RO^+ \\ \diagdown \\ R' \diagup \end{array} \ddot{C}:$$

$$\begin{array}{c} R_2N \\ \diagdown \\ R' \diagup \end{array} C: \quad \longleftrightarrow \quad \begin{array}{c} R_2N^+ \\ \diagdown \\ R' \diagup \end{array} \ddot{C}:$$

Dihalogenocarbenes show hardly any nucleophilic character, however; in their typical reactions they behave as electrophiles. Their stability increases in the order $:CBr_2 < :CCl_2 < :CF_2$.

Acylcarbenes and alkoxycarbonylcarbenes, which have triplet ground states, can react as 1,3-dipolar intermediates. Again, this behaviour can be rationalized in valence-bond terms by considering the resonance forms of the carbenes:

$$\begin{array}{c} R \\ \diagdown \\ C \\ \| \\ :O: \end{array} \ddot{C}H \quad \longleftrightarrow \quad \begin{array}{c} R \\ \diagdown \\ C \\ | \\ :\underset{\cdot\cdot}{O}: \end{array} \overset{+}{C}H$$

Substituents have an analogous effect on the structure of nitrenes. Most nitrenes have a triplet ground state, and are typically electrophilic in their reactions. Acylnitrenes and alkoxycarbonylnitrenes, like the corresponding carbenes, can act as 1,3-dipolar intermediates. Aminonitrenes and similar intermediates may have especially stable singlet states, like the corresponding carbenes:

$$R_2\ddot{N}-N: \quad \longleftrightarrow \quad R_2\overset{+}{N}=\ddot{N}:$$

Cyanonitrene[2] is probably unique in having a highly stabilized, symmetrical triplet ground state:

$$\cdot\ddot{N}=C=\ddot{N}\cdot$$

Not much is known about the relationship between structure and reactivity of arynes. Arynes can be formed across the adjacent carbon atoms in most positions of six-membered-ring aromatic compounds, and they have been postulated (but not proved) in smaller aromatic rings, such as five-membered-ring aromatic carbocycles and heterocycles. Intuitively, we might expect their relative stability to depend on the inherent stability of the parent aromatic system, and on the degree of overlap of the extra adjacent orbitals. The latter in turn should be related to the length of the bond in the original hydrocarbon: the shorter the bond, the greater the degree of overlap in the aryne, for a given ring size. However, it is very difficult to compare stabilities by chemical means. There have been attempts to do so by measuring the selectivity of arynes in the competitive addition of two nucleophiles, but the observed differences are small and probably depend more on the steric effects of the nucleophiles than on the selectivity of the arynes.[3] It has also been suggested that the stability of some hetarynes might be increased by overlap of the lone pair of electrons of the hetero atom with an adjacent sp^2 orbital, as in 2,3-dehydropyridine for example:[4]

The physical data available on arynes are meagre. Benzyne almost certainly has a singlet ground state, since benzyne generated and trapped in a frozen glass at 77°K gave no electron spin resonance signal.[5] The ultraviolet spectrum of benzyne has been recorded:[6] it shows a broad maximum centred at 242 nm. In these experiments, the lifetime of benzyne was about 0·0005 s, but at very low concentrations its lifetime can be as much as 0·02 s, as trapping experiments have shown.[7]

There are a few other intermediates related to arynes either by structure or by a similarity in their chemistry. The dehydro centres need not be adjacent in the aromatic ring: for example, 1,3- and 1,4-dehydrobenzene have been generated, 1,8-dehydronaphthalene has been generated and trapped, and 2,6-dehydropyridine has been suggested as an intermediate:

Medium- and small-ring cycloalkynes are another related group. They are not strictly analogous to arynes because they are true acetylenes, but are highly reactive because the triple bond is constrained in a small ring; however, their preparation and properties often resemble those of benzyne fairly closely. Cyclo-octyne is the smallest cycloalkyne that can be isolated; cycloheptyne, cyclohexyne and cyclopentyne are known only as reaction intermediates. Among

the more exotic cycloalkynes that have been postulated as intermediates are dehydrocyclo-octatetraene[8] [**10**] and dehydrobull-
valene[9] [**11**].

[**10**]   [**11**]

1–3   General methods of formation

The formal similarity between carbenes, nitrenes, and arynes can be
illustrated by considering the general methods available for generating them.   For all three types, the reaction involves an elimination,
to leave electrons distributed between two orbitals, which, in the
starting material, were both bonding orbitals:

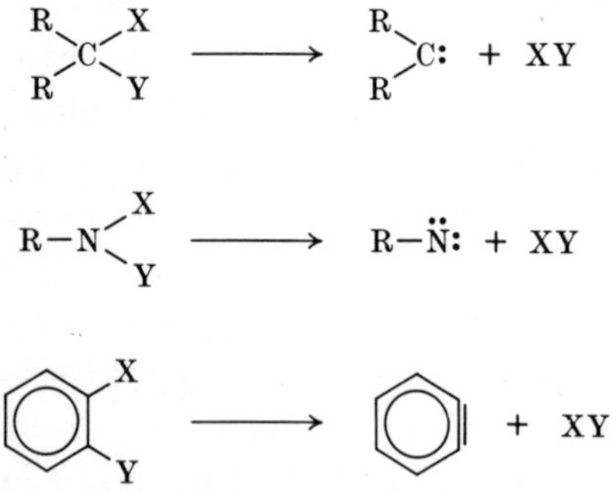

$$\begin{array}{c} R \\ \diagdown \\ R \diagup \end{array} C \begin{array}{c} X \\ \diagup \\ \diagdown Y \end{array} \longrightarrow \begin{array}{c} R \\ \diagdown \\ R \diagup \end{array} C: + XY$$

$$R-N \begin{array}{c} X \\ \diagdown Y \end{array} \longrightarrow R-\ddot{N}: + XY$$

For all three, X and Y may be independent atoms or groups, or
they may be joined as part of a cyclic system. For carbenes and
nitrenes there is also the possibility that X and Y can be replaced by
one doubly-bound group:

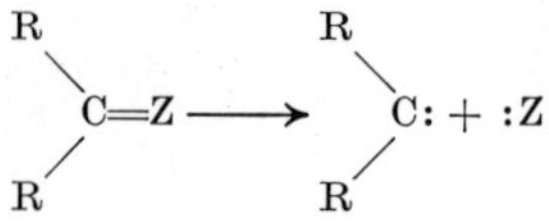

$$\begin{array}{c} R \\ \diagdown \\ R \diagup \end{array} C{=}Z \longrightarrow \begin{array}{c} R \\ \diagdown \\ R \diagup \end{array} C: + :Z$$

In all cases, the elimination may be concerted or stepwise. If it is
stepwise, the first bond cleavage may be homolytic, to give a radical
intermediate, or it may be heterolytic, to give a negatively charged
intermediate (a carbanion or nitrogen anion) or a positively charged
species (a carbonium ion or nitrenium ion). The possible intermediates
are shown opposite:

Carbene

$$\underset{R}{\overset{R}{>}}\ddot{C}-Y \qquad \underset{R}{\overset{R}{>}}\overset{-}{\ddot{C}}-Y \qquad \underset{R}{\overset{R}{>}}\overset{+}{C}-Y$$

Nitrene

$$\underset{R}{\overset{}{\diagdown}}\overset{\cdot}{N}-Y \qquad \underset{R}{\overset{}{\diagdown}}\overset{-}{\ddot{N}}-Y \qquad \underset{R}{\overset{}{\diagdown}}\overset{+}{N}-Y$$

Benzyne

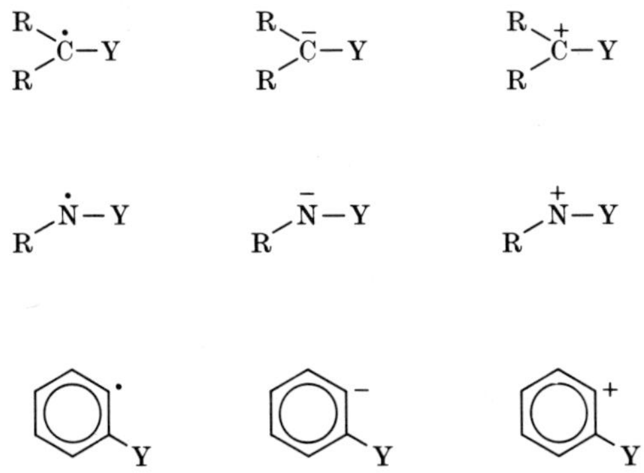

In practice, we find that the method of preparation of one type of the three intermediates can often be successfully adapted to the preparation of one or both of the other types. For example, the well-known decomposition of keten to methylene and carbon monoxide has found a fairly recent parallel in the formation of nitrenes from iso-cyanates,[10] and has an analogy in the formation of cycloheptyne from the corresponding cyclopropenone:[11]

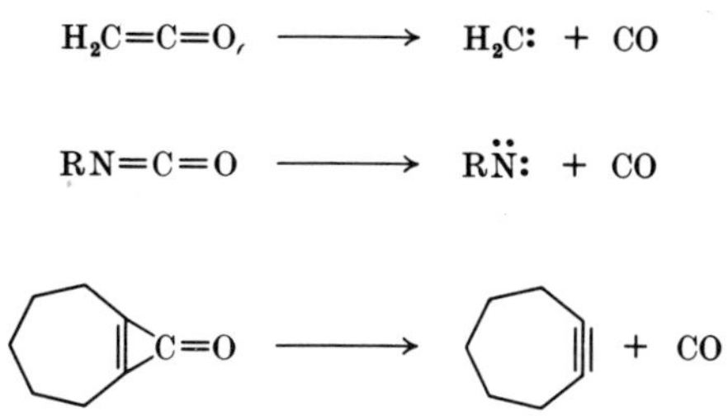

$$H_2C=C=O, \longrightarrow H_2C: + CO$$

$$RN=C=O \longrightarrow R\ddot{N}: + CO$$

+ CO

There is still plenty of scope for finding new ways of generating the intermediates, by applying the methods used for one type to the others. Indeed, the principles can be extended to inorganic analogues of carbenes and nitrenes. The chemistry of such intermediates has been reviewed.[12]

1–4   Evidence for the intermediates in reactions

The evidence that carbenes, nitrenes, or arynes are intermediates in a particular reaction has to be examined very critically. Only in very few cases has a detailed kinetic analysis been carried out that estab-lishes the presence of an intermediate. Usually the presence of the intermediate is postulated by analogy with other well-known re-actions, or because two independent sources of the intermediate lead to the same products. Most commonly of all, the intermediate is 'trapped' as a stable adduct: the cycloadditions of carbenes and nitrenes to olefins, and of arynes to dienes, are often used in this way:

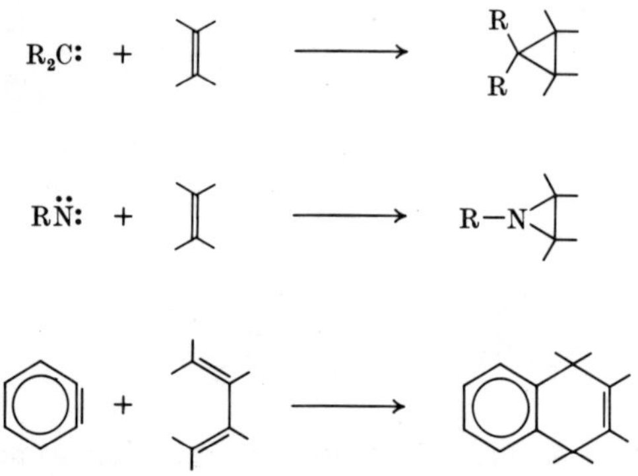

The isolation of the expected adducts is not, by itself, sufficient evidence for the existence of the intermediates in the reaction, however. Quite often, the precursor or some other reaction intermediate can react with the trap to give the expected adduct. For example, diazoalkanes are a well-known source of carbenes, but they can also react with olefins to give cyclopropanes by a route that does not involve carbenes:

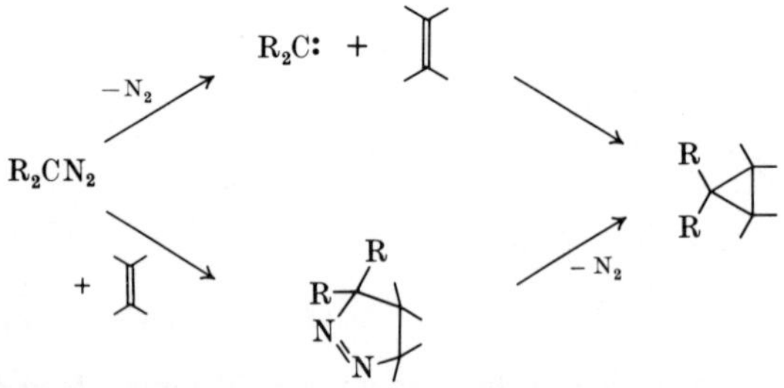

Such possibilities must be excluded before the trapping experiment can be taken as evidence for the presence of the intermediate.

References

1. R. Hoffmann, G. D. Zeiss, and G. W. Van Dine, *J. Amer. Chem. Soc.*, 1968, **90**, 1485.
2. A. G. Anastassiou and H. E. Simmons, *J. Amer. Chem. Soc.*, 1967, **89**, 3177.
3. Th. Kauffmann, H. Fischer, R. Nürnberg, M. Vestweber, and R. Wirthwein, *Tetrahedron Lett.*, 1967, 2911, 2917.
4. H. L. Jones and D. L. Beveridge, *Tetrahedron Lett.*, 1964, 1577.
5. R. W. Murray, unpublished (quoted by R. W. Hoffmann in *Dehydrobenzene and Cycloalkynes*).
6. M. E. Schafer and R. S. Berry, *J. Amer. Chem. Soc.*, 1965, **87**, 4497.

7. H. F. Ebel and R. W. Hoffmann, *Annalen*, 1964, **673**, 1.
8. A. Krebs and D. Byrd, *Annalen*, 1967, **707**, 66.
9. G. Schröder, H. Röttele, R. Merényi, and J. F. M. Oth, *Chem. Ber.*, 1967, **100**, 3527.
10. J. H. Boyer, W. E. Krueger, and G. J. Mikol, *J. Amer. Chem. Soc.*, 1967, **89**, 5504.
11. A. W. Krebs, *Angew. Chem. Internat. Edn.*, 1965, **4**, 10.
12. O. M. Nefedov and M. N. Manakov, *Angew. Chem. Internat. Edn.*, 1966, **5**, 1021.

2 Generation of carbenes

2–1 Introduction

In Chapter 1 the methods available for generating carbenes were outlined. In principle, carbenes can be formed by a concerted elimination, or via carbanion, radical, or carbonium ion intermediates. Methods of generating these intermediates may therefore also be potential methods of generating carbenes. Normally, carbanions, radicals, and carbonium ions react by some pathway that does not involve carbenes, so they must require some special structural features for carbenes to be generated from them. Although these structural requirements can quite often be met in carbanions and radicals, they are very rarely found in carbonium ions, which are therefore not normally sources of carbenes.

The mechanisms by which carbenes are generated from many precursors have not been fully investigated, so it is not possible to adopt a rigid classification of the methods based on mechanism. However, we can recognize similarities in the structures of several of the precursors and the methods used to produce carbenes from them, and these enable us to make some correlations. For example, diazoalkanes are the best known sources of many classes of carbenes; ketens, which show certain structural similarities to diazoalkanes, can also give carbenes in similar conditions. Sulphur ylids and phosphorus ylids, which are also structurally related to diazoalkanes, have occasionally been found to give carbenes.

$$R_2C{=}N_2 \longleftrightarrow R_2\overset{-}{C}{-}\overset{+}{N_2} \longrightarrow {:}CR_2 + N_2$$

$$R_2C{=}C{=}O \longleftrightarrow R_2\overset{-}{C}{-}\overset{+}{C}{=}O \longrightarrow {:}CR_2 + CO$$

$$R_2C{=}PR'_3 \longleftrightarrow R_2\overset{-}{C}{-}\overset{+}{P}R'_3 \longrightarrow {:}CR_2 + PR'_3$$

There are also a large number of routes to carbenes involving halogeno compounds with bases or metals. In many of these, it is doubtful whether 'free' carbenes are formed at all; the carbene may be complexed with a metal or held in a solvent 'cage' with a salt, or the intermediate may be an organometallic compound rather than a carbene. Organometallic, carbene-like intermediates are usually called *carbenoids*.[1] It is not yet possible to define clearly which of these reactions are likely to produce free carbenes, and which are not. As the organometallic routes are often very useful synthetic procedures, we have included them in this chapter.

2–2   Carbenes from diazoalkanes, ketens, and ylids

From diazoalkanes. The photolysis or thermolysis of diazoalkanes provides the most common general route to carbenes. The first suggestions that the decomposition of diazomethane might involve methylene, $:CH_2$, as an intermediate were made by Nef at the end of the last century. Quite recently, the visible and ultraviolet spectra of singlet and triplet methylene have been observed directly from the flash photolysis of diazomethane.   Carbenes have also been detected by electron spin resonance when other diazoalkanes have been photolysed in a matrix at very low temperatures; diphenyldiazomethane and diazofluorene are examples. It therefore seems that the formation of carbenes by the photolysis of diazoalkanes is quite general, at least when the reaction is carried out in an aprotic solvent. In a protic solvent, such as an alcohol, it is likely that the mode of decomposition of the diazoalkane involves a prior protonation, so that a carbonium ion, rather than a carbene, is formed when nitrogen is lost:

$$R_2C{=}\overset{+}{N}{=}\overset{-}{N} \longleftrightarrow R_2\overset{-}{C}{-}\overset{+}{N}{\equiv}N \xrightarrow{\ H^+\ } R_2CH{-}\overset{+}{N}{\equiv}N \longrightarrow R_2\overset{+}{C}H + N_2$$

Carbenes formed by photolysis of diazoalkanes are highly energetic species, and their reactions may be indiscriminate. For example, the photolysis of diazomethane produces methylene, which can insert into primary, secondary, and tertiary C–H bonds of an alkane with almost equal ease, as well as adding to double bonds. For this reason, photolysis of the diazoalkane is often not a good way of generating the carbene for some synthetic purpose involving another molecule. Thermal decomposition may produce a less energetic carbene, but it has the disadvantage that other modes of reaction of the diazoalkane, not involving carbenes, become important: the addition of diazoalkanes to olefins, to form pyrazolines, for example. Also, some diazoalkanes are rather stable thermally, and the reaction temperature may therefore be high. Diazoesters and the resonance-stabilized diazocyclopentadiene derivatives [1] are in this category.

[1]

Copper powder or copper salts lower the decomposition temperature of the diazoalkanes and give intermediates that are very selective in their reactions: the insertion into C–H bonds of other molecules is usually suppressed, for example. It is doubtful whether these intermediates are 'free' carbenes, however; copper–carbene complexes seem much more likely. These catalysed diazoalkane decompositions therefore properly belong to the 'carbenoid' class.

If the diazoalkanes are difficult or dangerous to handle, it is usually possible to obtain the carbenes directly from a precursor of the diazoalkanes. Three groups of compounds that have been used to generate carbenes by this means are hydrazones [**2**], toluene-*p*-sulphonylhydrazones [**3**], and *N*-nitrosoalkylureas [**4**].

$$R_2C=N-NH_2 \qquad\qquad R_2C=N-N\overset{H}{\underset{SO_2}{\diagdown}}\!\!\!-\!\!\!\langle\bigcirc\rangle\!-Me \qquad\qquad R_2CH-N\overset{NO}{\underset{CONH_2}{\diagdown}}$$

[**2**]   [**3**]   [**4**]

From hydrazones. Hydrazones can be oxidized to diazoalkanes in mild conditions. The oxidizing agent is usually mercuric oxide, but others, such as silver oxide and lead tetra-acetate, can also be used. The diazoalkanes need not be isolated and can be decomposed directly to the carbenes:

$$R_2C{=}N{-}NH_2 \longrightarrow R_2C{=}N_2 \longrightarrow R_2C{:} + N_2$$

This route is often used for dialkyl derivatives of diazomethane.

From toluene-p-sulphonylhydrazones. The toluene-*p*-sulphonylhydrazones are converted into their sodium or lithium salts, which are then pyrolysed or photolysed. In these conditions, the diazoalkanes cannot be isolated and the carbenes are generated directly. A typical procedure would involve heating the toluene-*p*-sulphonylhydrazone with an equimolar amount of sodium methoxide in an inert aprotic solvent at 150–200°:

$$R_2C{=}N{-}NHSO_2\!\!-\!\!\langle\bigcirc\rangle\!-Me \xrightarrow[\text{or LiBu}]{\text{NaOMe}} R_2C{=}N{-}\bar{N}SO_2\!\!-\!\!\langle\bigcirc\rangle\!-Me$$

$$\longrightarrow \bar{S}O_2\!\!-\!\!\langle\bigcirc\rangle\!-Me \;+\; {:}CR_2 \;+\; N_2$$

Alternatively, the sodium or lithium salt can be isolated and then thermolysed[2] or photolysed.[3]

Among the wide range of carbenes generated by this method are many with alkyl, cycloalkyl, and alkoxy substituents. Some examples are shown:

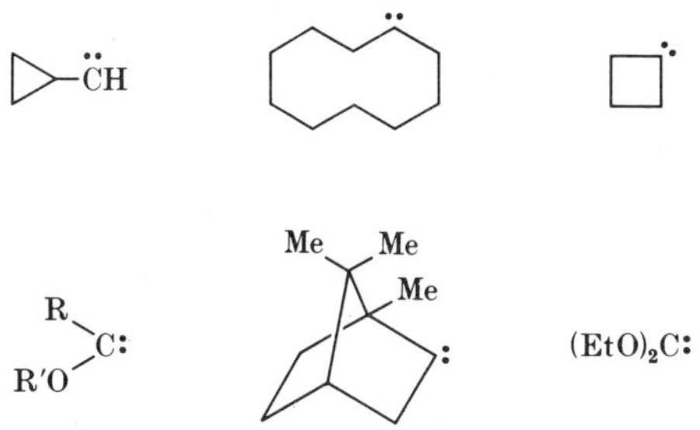

From N-*nitrosoalkylureas.* The reaction of N-nitrosoalkylureas with bases is a means of making diazoalkanes, and it has also been used to generate carbenes directly, with lithium ethoxide as the base and a low reaction temperature:[4]

$$R_2CHNCONH_2 \xrightarrow{\text{base}} R_2CN_2 \longrightarrow R_2C: $$
$$\underset{NO}{|}$$

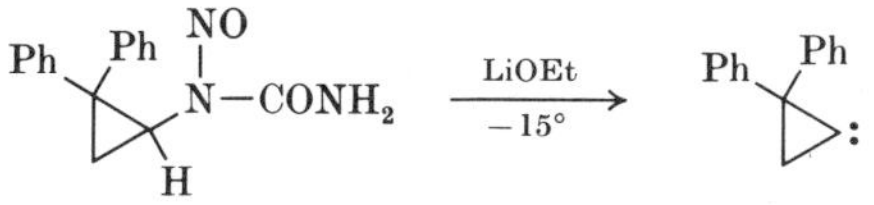

From ketens. Like diazoalkanes, ketens can be sources of carbenes, since the reaction:

$$R_2C{=}C{=}O \longrightarrow R_2C: + CO$$

involves the loss of a stable molecule, carbon monoxide.

There is good evidence (from an examination of the reaction products) that keten does give methylene on pyrolysis or photolysis. Photolysis of keten with light of wavelength below 280 nm is believed to give methylene by direct dissociation, but light of wavelength greater than 280 nm probably produces decomposition via an excited keten molecule, since added oxygen inhibits the reaction. Methylene generated from keten, like that from diazomethane, is a very indiscriminate reagent and therefore not much used in preparative chemistry.

Substituted ketens can similarly give carbenes by thermolysis or photolysis; for example, diphenylcarbene can be generated from the keten:

$$Ph_2C{=}C{=}O \longrightarrow Ph_2C{:} + CO$$

This is not a good general route to substituted carbenes, however, because the ketens tend to polymerize in the conditions needed to generate the carbenes.

From ylids. Sulphur ylids, phosphorus ylids, and nitrogen ylids can react with olefins to give cyclopropanes, but carbenes are not usually intermediates. The ylid acts as a nucleophile, attacking an electrophilic double bond, for example:[5]

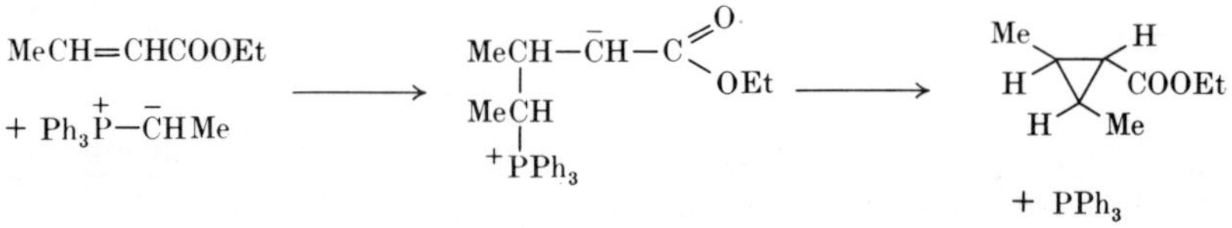

MeCH=CHCOOEt

+ $Ph_3\overset{+}{P}{-}\overset{-}{C}HMe$

$\longrightarrow$

$MeCH{-}\overset{-}{C}H{-}C\overset{\nearrow O}{\diagdown OEt}$
$\quad\;\, |$
$\quad MeCH$
$\quad\;\, |$
$\quad\; \overset{+}{P}Ph_3$

$\longrightarrow$

Me, H / H, COOEt / H, Me (cyclopropane)

+ PPh_3

This is only a minor reaction pathway in some systems: with $\alpha\beta$-unsaturated ketones, for example, there may be preferential addition to the carbonyl double bond with the formation of an epoxide.

There are a few reactions in which ylids give cyclopropanes with nucleophilic olefins, and, in these, carbenes may be intermediates. The production of formal 'carbene dimers' in ylid reactions can also be explained by a carbene mechanism. Examples of such reactions involving sulphur ylids,[6] phosphorus ylids,[7] and nitrogen ylids[8] are shown:

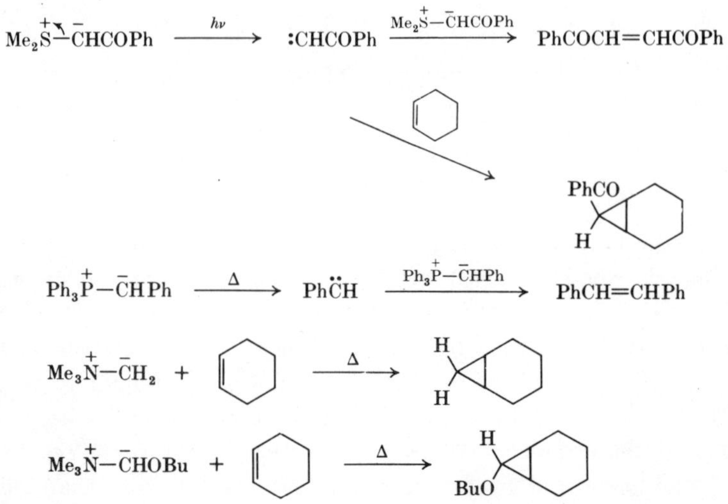

$Me_2\overset{+}{S}{-}\overset{-}{C}HCOPh \xrightarrow{\ h\nu\ } {:}CHCOPh \xrightarrow{\ Me_2\overset{+}{S}{-}\overset{-}{C}HCOPh\ } PhCOCH{=}CHCOPh$

$Ph_3\overset{+}{P}{-}\overset{-}{C}HPh \xrightarrow{\ \Delta\ } Ph\overset{..}{C}H \xrightarrow{\ Ph_3\overset{+}{P}{-}\overset{-}{C}HPh\ } PhCH{=}CHPh$

$Me_3\overset{+}{N}{-}\overset{-}{C}H_2 + $ (cyclohexene) $\xrightarrow{\ \Delta\ }$ (bicyclic, H H)

$Me_3\overset{+}{N}{-}\overset{-}{C}HOBu + $ (cyclohexene) $\xrightarrow{\ \Delta\ }$ (bicyclic, H BuO)

endo and *exo*

2–3   Carbenes by other fragmentations

There is a growing number of routes to carbenes that are related in that the carbenes are produced from the precursor simply by photolysis (or, occasionally, by thermolysis). These reactions may be concerted, or they may involve radical intermediates. Two factors seem to govern the feasibility of the reaction: the precursor must be sterically strained (that is, it must have high ground-state energy) and the non-carbene fragment that is produced in the reaction (if there is one) must be thermodynamically stable, so that there is low transition-state energy. Precursors that can meet these requirements include certain oxiranes (epoxides), diazirines, cyclopropanes, and sterically crowded olefins.

From oxiranes. Phenyl-substituted oxiranes can give phenyl-substituted carbenes in good yield when they are photolysed.[9] This is a useful way of generating phenylcarbene and phenylcyanocarbene:

$$\text{Ph} \underset{\text{H}}{\overset{\text{O}}{\triangle}} \text{H} \quad \xrightarrow{h\nu} \quad \text{Ph}\ddot{\text{C}}\text{H} \; + \; \text{PhCHO}$$

$$\text{Ph} \underset{\text{NC}}{\overset{\text{O}}{\triangle}} \text{Ph} \quad \xrightarrow{h\nu} \quad \text{Ph}\ddot{\text{C}}\text{CN} \; + \; \text{Ph}_2\text{CO}$$

From diazirines. The photolysis of alkyl-substituted diazirines is a well-established way of generating alkylcarbenes, and is useful because the diazirines tend to be less hazardous to handle than the isomeric diazoalkanes. The diazirines can be synthesized from a ketone, ammonia, and chloramine, followed by oxidation of the diaziridines:

$$\text{R}_2\text{C}{=}\text{O} + \text{NH}_3 + \text{NH}_2\text{Cl} \longrightarrow \text{R}_2\text{C}\underset{\text{NH}}{\overset{\text{NH}}{\big\langle}} \longrightarrow \text{R}_2\text{C}\underset{\text{N}}{\overset{\text{N}}{\big\langle}}$$

Halogenodiazirines have also been used to generate halogenocarbenes, many of which are not available by other routes.[10] The method also has the advantage that the reaction conditions are neutral, whereas most methods of generating halogenocarbenes require base. Carbenes made this way include $\ddot{\text{C}}\text{F}_2$, Me$\ddot{\text{C}}Cl, MeO\ddot{\text{C}}$Cl, and Ph$\ddot{\text{C}}$Br.

From cyclopropanes. A few cyclopropanes have been shown to generate carbenes on photolysis. 9,10-Dihydro-9,10-methanophenanthrene [5],[11] phenylcyclopropane,[11] and 1,1-dichloro-2-phenylcyclopropane[12] are examples:

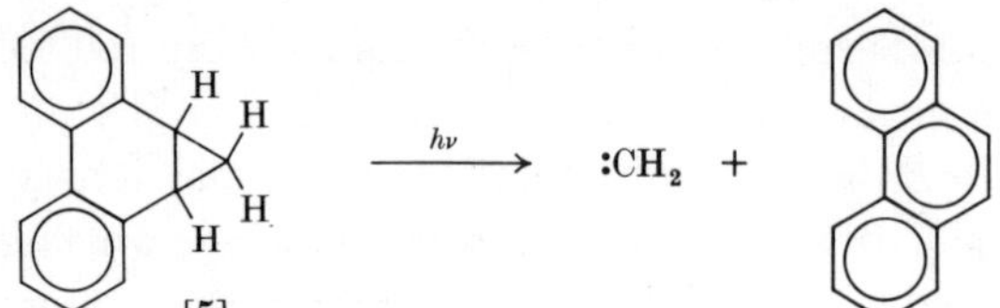

$$[5] \xrightarrow{h\nu} :CH_2 +$$

$$\xrightarrow{h\nu} :CH_2 + PhCH{=}CH_2$$

$$\xrightarrow{h\nu} :CCl_2 + PhCH{=}CH_2$$

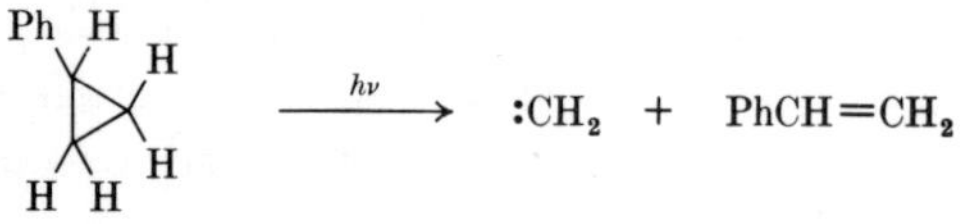

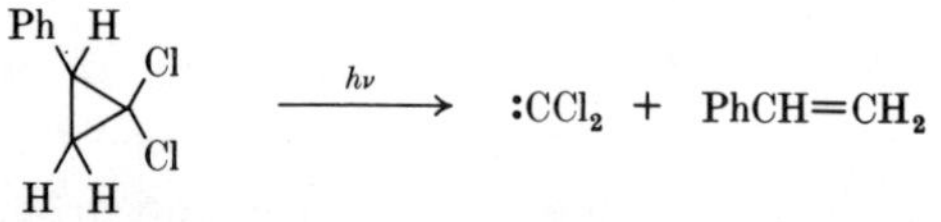

Similarly, hexafluorocyclopropane can give difluorocarbene on thermolysis :[13]

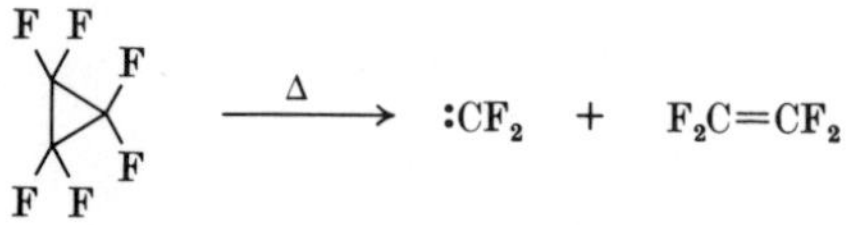

$$\xrightarrow{\Delta} :CF_2 + F_2C{=}CF_2$$

From olefins. Some sterically crowded olefins, such as [6], are thought to dissociate on heating to give carbene monomers. The ground-state energy is likely to be high in such olefins, with a considerable distortion from a planar structure.

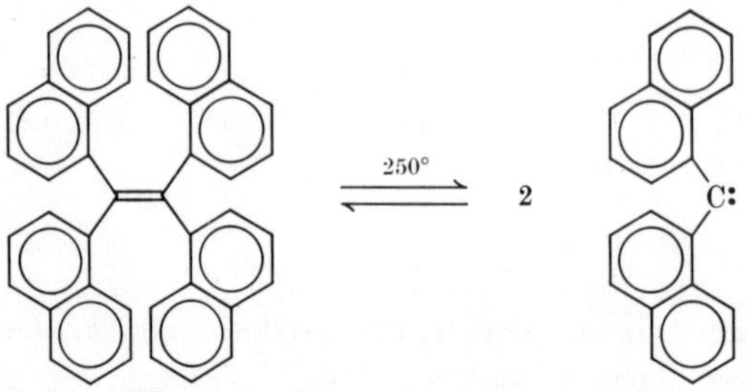

$$[6] \underset{250°}{\rightleftharpoons} 2$$

From miscellaneous precursors. If the conditions are vigorous enough, even stable molecules like methane, chloroform, and carbon tetrachloride can give carbenes. Methylene is produced by the vacuum ultraviolet photolysis of methane, and dichlorocarbene from the pyrolysis of chloroform or carbon tetrachloride at 1500°.[14] Di-iodomethane can be photolysed to produce methylene in much milder conditions.[15] The production of atomic carbon in a low-intensity carbon arc under high vacuum is, in a sense, a fragmentation reaction. Skell has shown that the atoms can behave as bis-carbenes, and their reactions with olefins constitute a method of generating monocarbenes. Thus, for example, a carbon atom with an olefin will produce a cyclopropylidene as an intermediate:[16]

Other extrusion reactions that can produce carbenes are those of norbornadienone ketals[17] and of bridged 10-annulenes:[18]

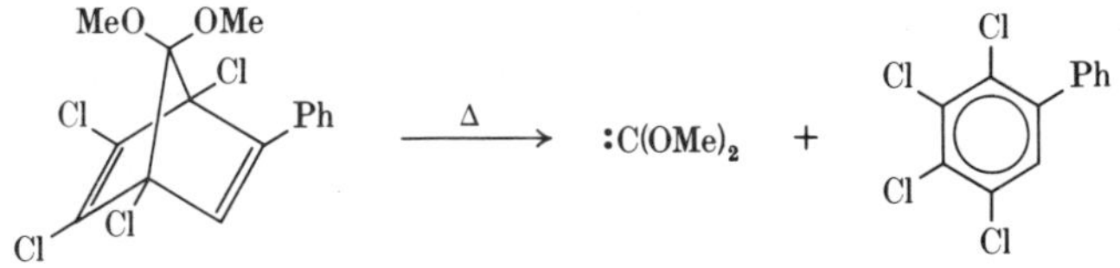

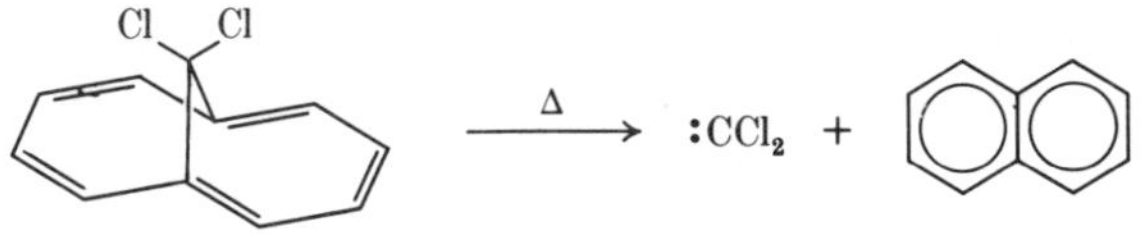

2–4   Carbenes from carbanions

The modern development of carbene chemistry started in the 1950s with the elucidation of the mechanism of the basic hydrolysis of chloroform, by J. Hine and his co-workers. They showed that the trichloromethyl anion and dichlorocarbene were both intermediates in the reaction, the carbene being formed from the carbanion in the rate-determining step:

$$^-CCl_3 \longrightarrow :CCl_2 + Cl^-$$

Some of the evidence on which this conclusion is based is as follows. In deuterated solvents, the basic hydrolysis of chloroform is slow compared with the rate of incorporation of deuterium into the chloroform, so there must be a rapid pre-equilibrium:

$$CHCl_3 + {}^-OH \rightleftharpoons {}^-CCl_3 + H_2O$$

in the hydrolysis by hydroxide ions. The subsequent hydrolysis could go by one of two mechanisms, outlined below:

(a) $H_2O + {}^-CCl_3 \xrightarrow{slow} Cl^- + H_2\overset{+}{O}{-}\overset{-}{C}Cl_2 \xrightarrow{fast} CO, HCO_2^-$, etc

(b) $\quad\quad {}^-CCl_3 \xrightarrow{slow} Cl^- + :CCl_2 \xrightarrow[H_2O]{fast} CO, HCO_2^-$, etc

The major evidence that (*b*) is the correct mechanism comes from an investigation of the effect of added salts on the reaction rate. The anions F^-, NO_3^-, and ClO_4^- do not affect the rate of chloroform hydrolysis, but I^-, Br^-, and Cl^- all decrease it, the magnitude of the effect being in the order of the nucleophilicities ($I^- > Br^- > Cl^-$). If mechanism (*a*) were correct, these anions should hardly affect the rate, since a competition with water for the trichloromethyl anion would simply generate another trihalomethyl anion:

$$X^- + {}^-CCl_3 \longrightarrow Cl^- + {}^-CXCl_2$$

According to mechanism (*b*), however, the effect of the added anions should be to slow down the reaction in the order found, since their reaction with dichlorocarbene will constitute a reversal of the rate-determining step.

This method of generating dichlorocarbene has been developed and extended both to other haloforms and to other precursors of tri-halogenomethyl anions. A major advance was the discovery that, if the reaction was carried out in a non-aqueous solvent, by using a strong base that was also a rather poor nucleophile, the carbenes could be trapped by olefins to give cyclopropanes. Solvents that are commonly used include benzene, toluene, 1,2-dimethoxyethane, and dimethyl sulphoxide, and potassium t-butoxide is the usual base. The relative ease of loss of the halogens as anions is $I \sim Br > Cl \gg F$, so that $CHBr_2F$ gives the carbene $:CBrF$, $CHBrClF$ gives $:CClF$, $CHClF_2$ gives $:CF_2$, and so on. For the haloforms CHF_2X ($X = Br$, Cl, I) it is likely that the elimination of HX to give difluorocarbene is concerted and does not involve an intermediate anion.

This method of generating dihalogenocarbenes is widely used in organic synthesis. Its disadvantages are the moderate yields of carbene adducts obtained, these being partly a consequence of the necessary formation of alcohol in the reaction:

$$CHX_3 + {}^-OR \rightleftharpoons {}^-CX_3 + ROH$$
$$\downarrow \qquad \qquad |$$
$$:CX_2 \longrightarrow \text{side reactions}$$

and the use of strong base, which limits the reactions to substrates that are not base-sensitive. Other sources of trihalogenomethyl anions have been found that avoid one or both of these disadvantages. Thus, ethyl trichloroacetate reacts with alkoxide ions to give tri-chloromethyl anions. Since this reaction does not involve an alcohol as a by-product, carbene adducts can be obtained in very good yields:

$$CCl_3COOEt + {}^-OR \rightleftharpoons Cl_3C\overset{\displaystyle O^-}{\underset{\displaystyle OEt}{\vert\!-\!C\!-\!}}OR \rightleftharpoons {}^-CCl_3 + O{=}C\overset{\displaystyle OR}{\underset{\displaystyle OEt}{\diagup}}$$

$$\downarrow$$

$$:CCl_2 + Cl^-$$

Ethyl chlorodifluoroacetate has been used similarly as a source of difluorocarbene. Trichloromethylsulphinic acid esters and hexachloroacetone can also be used as sources of dichlorocarbene:

$$CCl_3SO_2Me + {}^-OR \rightleftharpoons \bar{C}Cl_3 + ROSO_2Me$$

$$CCl_3COCCl_3 + 2{}^-OR \longrightarrow 2\bar{C}Cl_3 + (RO)_2CO$$

A related method that avoids the use of an external base is the thermolysis of the sodium salt of trichloroacetic acid (or of tribromoacetic acid) in an aprotic solvent, such as 1,2-dimethoxyethane. This route can give very high yields of carbene adducts with olefins:

$$CCl_3COO^-Na^+ \longrightarrow {}^-CCl_3 + CO_2 + Na^+$$

$$\downarrow$$

$$:CCl_2 + Cl^-$$

A very good route to dihalogenocarbenes, developed by Seyferth, involves the thermolysis of phenylmercury trihalogenomethanes, $PhHgCX_3$. These reactions may involve trihalogenomethyl anions as intermediates. Even with deactivated olefins, good yields of the dihalogenocyclopropanes can often be obtained by this route:

$$PhHgCCl_2Br \xrightarrow{80°} PhHgBr + :CCl_2$$

$$PhHgCBr_3 \longrightarrow PhHgBr + :CBr_2$$

In a somewhat similar reaction, trimethyltrifluoromethyltin, Me_3SnCF_3, gives difluorocarbene when heated in an inert solvent with sodium iodide:

$$Me_3SnCF_3 + NaI \longrightarrow Me_3SnI + :CF_2 + NaF$$

Attempts to form carbenes other than dihalogenocarbenes through intermediate carbanions have been less successful. Dichloromethane, for example, can give monochlorocarbene with potassium t-butoxide in a reaction analogous to that of chloroform, but the yields of carbene adducts are poor:

$$CH_2Cl_2 \xrightarrow{^{-}OBu^t} :CHCl + Bu^tOH + Cl^-$$

Better yields can be obtained by using n-butyl-lithium as the base, but probably dichloromethyl-lithium is the intermediate responsible for the formation of the adducts:

$$CH_2Cl_2 + nBu-Li \longrightarrow LiCHCl_2 \dashrightarrow LiCl + :CHCl$$

This exemplifies the problem with a wide range of dehydrohalogenations by strong bases such as sodamide, butoxide ions, sodium alkyls, and lithium alkyls; the products can be accounted for either by a carbene or by a carbenoid mechanism. When alkoxide ions are used for the dehydrohalogenations it is reasonable to postulate free carbenes as intermediates, by analogy with the haloform reactions, but with metal alkyls as the bases free carbene intermediates are much less likely. In terms of products obtained, however, the nature of the base often makes little difference. For example, a good general route to cyclopropenes is the dehydrohalogenation of allyl halides, which can be carried out by using lithium alkyls, sodamide or potassium t-butoxide as the base. Alkenylcarbenes are possible intermediates. The pyrolysis of salts of toluene-p-sulphonylhydrazones of $\alpha\beta$-unsaturated ketones (Section 2–2) provides an independent route to the carbenes and leads to the same products as the dehydrohalogenation route:[19]

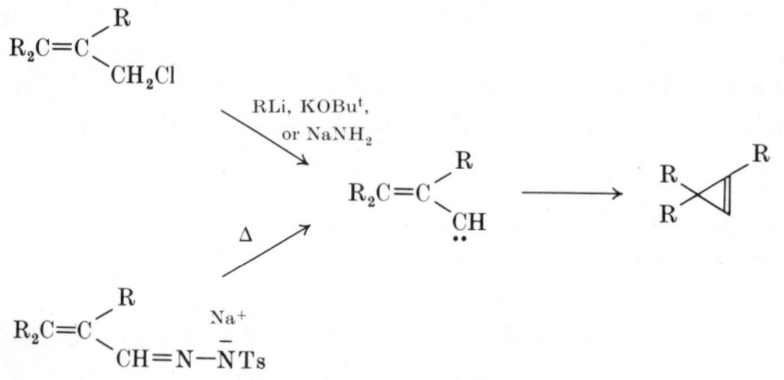

Benzyl halides and benzal halides can similarly be dehydrohalogenated and the products are formally derived from intermediate carbenes:

$$PhCHBr_2 \xrightarrow{^{-}OBu^t} Ph\ddot{C}Br + Br^- + Bu^tOH$$

$$PhCH_2Cl \xrightarrow{BuLi} Ph\ddot{C}HClLi \dashrightarrow Ph\ddot{C}H + LiCl$$

In the dehydrohalogenation of benzyl chloride by sodamide, the product, stilbene, is formed through an intermediate carbanion and carbenes are not involved:

$$PhCH_2Cl \xrightarrow{NaNH_2} Ph\bar{C}HCl \xrightarrow{PhCH_2Cl} PhCHClCH_2Ph + Cl^-$$

$$\downarrow$$

$$PhCH{=\!=}CHPh$$

1-Halogenoalkenes are another group of compounds that may or may not give carbene intermediates on dehydrohalogenation. If the substituents on C-2 are alkyl groups, the carbenes can be trapped with olefins:

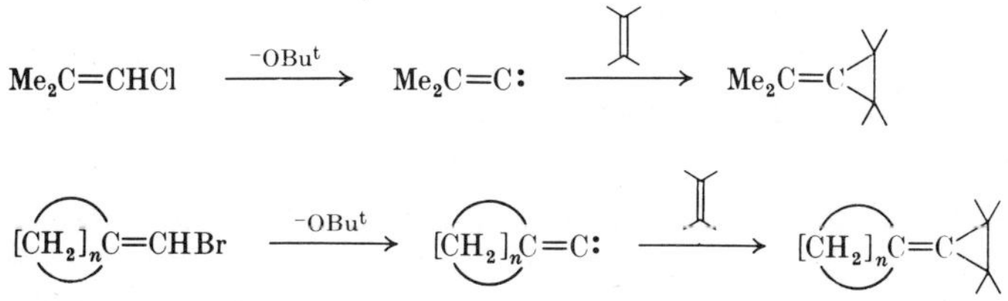

In the absence of a trap, the carbenes may rearrange to acetylenes.[20]

If the 2-substituents are hydrogen atoms or aryl groups, however, the carbenes cannot be trapped. The intermediate carbanion rearranges to an acetylene, with simultaneous loss of the halide ion. It is the group *trans* to the departing halogen that usually migrates:[1]

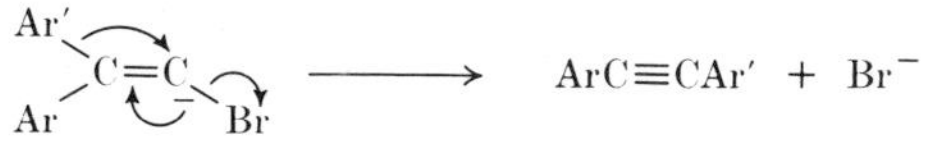

Allenylcarbenes can be obtained in a similar reaction from 1-bromoallenes and alkoxide ions. The isomeric 3-bromo-1-alkynes give the same carbenes. Both reactions involve a rapid reversible removal of the terminal protons followed by a rate-determining loss of bromide ions:[21]

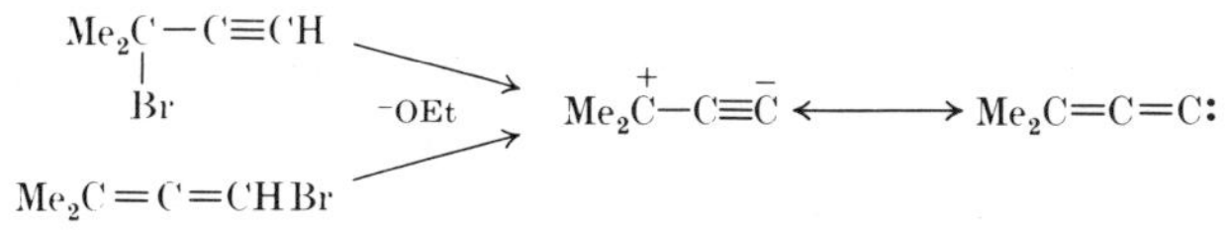

The anions of α-halogenoketones and α-halogenoesters are not carbene precursors. In basic conditions other reactions, such as the Favorskii rearrangement of α-halogenoketones and condensation reactions of halogenoesters, are more favourable. Similarly, salts of α-bromoacrylic acids are not carbene precursors.[22]

$$R_2C\!=\!C\!\!\begin{array}{c}Br\\ \diagup\\ \diagdown\\ COO^-\end{array} \xrightarrow{\;\Delta\;} R_2C\!=\!C^- \;\; -/\!/\!\!\rightarrow R_2C\!=\!C\!:$$

2–5   Carbenoids

gem-Dihalogeno compounds can be dehalogenated by metals or by metal alkyls, and the products obtained are often typical of those from carbenes: cyclopropanes with olefins, for example. Whenever the mechanism of one of these reactions has been investigated carefully, however, the evidence is nearly always against free carbenes as intermediates. In a reaction involving the formation of a cyclopropane from an olefin, for example, it may be found that the rate of the reaction is directly dependent on the olefin concentration: thus the olefin is involved in the rate-determining step of the reaction and free carbenes cannot be the reactive intermediates. Sometimes it has been possible to obtain the organometallic carbenoid intermediates in a stable form in solution and to study their reactions more directly: the elegant work of Köbrich[1] on α-halogenolithium compounds is an example.

   α-Halogeno-organolithium compounds. These compounds can be prepared from *gem*-dihalogeno compounds, by halogen–lithium exchange or by the replacement of an α-hydrogen of a halogeno compound by lithium from a lithium alkyl:

$$R_2CX_2 + Li \longrightarrow R_2C\!\!\begin{array}{c}Li\\ \diagup\\ \diagdown\\ X\end{array} \longleftarrow R'Li + R_2CHX$$

With *gem*-dibromo compounds, bromine is replaced by lithium in preference to hydrogen, but with *gem*-dichloro compounds an α-hydrogen may be preferentially displaced:

$$RCHBr_2 + R'Li \longrightarrow RCHBrLi + R'Br$$
$$RCHCl_2 + R'Li \longrightarrow RCCl_2Li + R'H$$

Köbrich has found that these α-halogenolithium derivatives are often stable in tetrahydrofuran solution at very low temperatures and can be detected, for example by carboxylation to give the corresponding carboxylic acids:

$$R_2CLiX + CO_2 \longrightarrow R_2CXCOO^- Li^+$$

These intermediates can act as electrophiles and their pattern of reactivity towards nucleophiles is broadly similar to that of the corresponding carbenes. How much carbene character the inter-mediates have in these reactions probably depends on the sub-stituents and the reaction conditions, but evidence for the existence of free carbenes is rare,[23] whereas that against free carbenes is common.[1]

Organozinc and related intermediates. Di-iodomethane reacts with zinc–copper couple in ether, or with zinc alone in dimethoxyethane, to give a 1:1 adduct, stable in solution, the structure of which is probably a loose association of solvated units of $IZnCH_2I$. The re-agent adds to nucleophilic olefins to give cyclopropanes, and the rate of its decomposition can be increased by the addition of olefins, show-ing that free methylene is not an intermediate. Similar 'methylene transfer' reagents, $BrZnCH_2Br$ and $ClZnCH_2Cl$, can be made by the reaction of diazomethane with zinc bromide or zinc chloride. Deriva-tives of other metals can also be used; for example, iodomethyl-magnesium iodide, ICH_2MgI, and chloromethyldiethylaluminium, Et_2AlCH_2Cl, can convert olefins into cyclopropanes in good yield. The reaction cannot be extended satisfactorily to other carbene trans-fers: although homologues of iodomethylzinc iodide have been made, they form little or no cyclopropanes with olefins.[24]

2–6   Other carbene-forming reactions

Although most of the important methods of generating carbenes have been outlined in the previous sections, there are a few other reaction types that have occasionally been used. Some examples are given below.

From carbonium ions. The formation of carbenes from carbonium ions, though feasible, is very uncommon. Two reactions that probably go through carbenes formed from carbonium ions are shown: the first is the aprotic diazotization of 9-(aminomethylene)fluorene [7] and the subsequent trapping of the carbene with cyclohexene;[25] the second, the reaction of 1,2-diphenylcyclopropene-3-carboxylic acid [8] with triphenylmethyl perchlorate.[26]

Rearrangement. Benzocyclobutenedione [9] rearranges photo-chemically to a carbene, which can be trapped by olefins and acetylenes.[27]

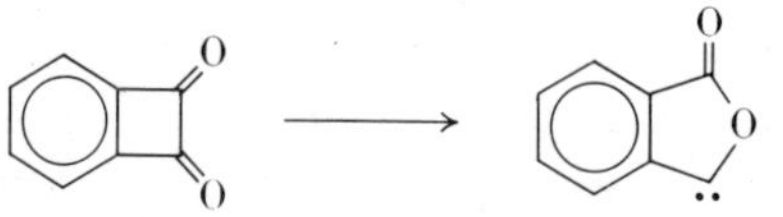

[9]

Deoxygenation of carbonyl compounds. Cyclic carbonates[28] and anhydrides[29] can be deoxygenated by tervalent phosphorus compounds to give products that might be formed through carbene intermediates, though the mechanism remains in doubt. A stereospecific olefin synthesis developed by Corey[30] involves a similar desulphurization of cyclic thionocarbonates.

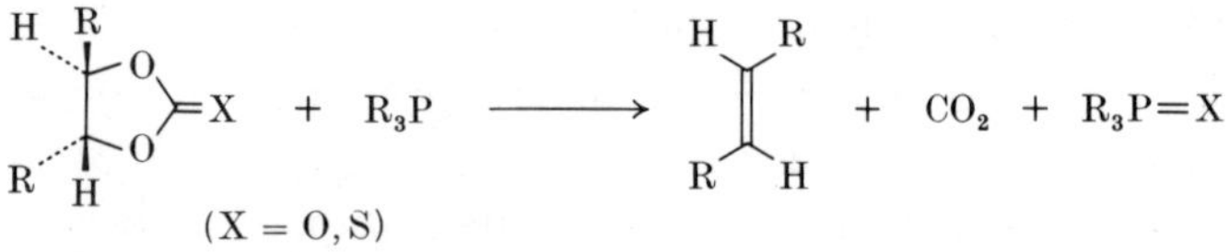

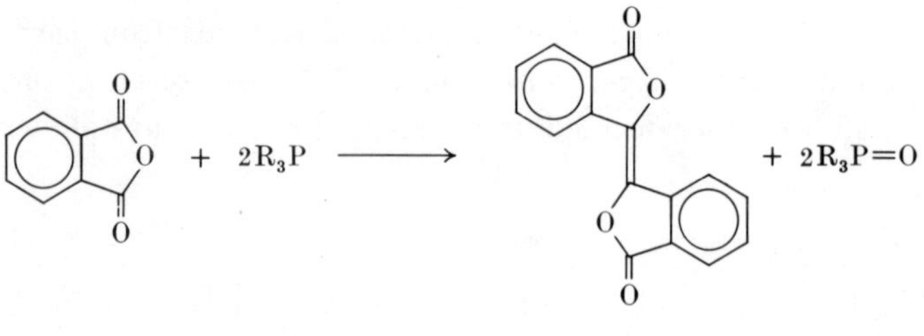

$(X = O, S)$

References

1. G. Köbrich, *Angew. Chem. Internat. Edn.*, 1967, **6**, 41.
2. R. M. McDonald and R. A. Krueger, *J. Org. Chem.*, 1966, **31**, 488.
3. W. M. Jones and C. L. Ennis, *J. Amer. Chem. Soc.*, 1967, **89**, 3069.
4. W. M. Jones, M. H. Grasley, and W. S. Brey, *J. Amer. Chem. Soc.*, 1963, **85**, 2754.
5. H. J. Bestmann and F. Seng, *Angew. Chem. Internat. Edn.*, 1962, **1**, 116.
6. B. M. Trost, *J. Amer. Chem. Soc.*, 1966, **88**, 1587.
7. S. Trippett, *Proc. Chem. Soc.*, 1963, 19.
   See also H. Tschesche, *Chem. Ber.*, 1965, **98**, 3318; A. Ritter and B. Kim, *Tetrahedron Lett.*, 1968, 3449.

8. G. Wittig and D. Krauss, *Annalen*, 1964, **679**, 34.
9. P. C. Petrellis, H. Dietrich, E. Meyer, and G. W. Griffin, *J. Amer. Chem. Soc.*, 1967, **89**, 1967.
10. W. H. Graham, *J. Amer. Chem. Soc.*, 1965, **87**, 4396.
11. D. B. Richardson, *et al.*, *J. Amer. Chem. Soc.*, 1965, **87**, 2763.
12. M. Jones, W. H. Sachs, A. Kulczycki, and F. J. Waller, *J. Amer. Chem. Soc.*, 1966, **88**, 3167.
13. J. M. Birchall, R. N. Haszeldine, and D. W. Roberts, *Chem. Comm.*, 1967, 287.
14. L. D. Wescott and P. S. Skell, *J. Amer. Chem. Soc.*, 1965, **87**, 1721.
15. D. C. Blomstrom, K. Herbig, and H. E. Simmons, *J. Org. Chem.*, 1965, **30**, 959.
16. P. S. Skell and R. R. Engel, *J. Amer. Chem. Soc.*, 1967, **89**, 2912.
17. R. W. Hoffmann and C. Wünsche, *Chem. Ber.*, 1967, **100**, 943.
18. V. Rautenstrauch, H. J. Scholl, and E. Vogel, *Angew. Chem. Internat. Edn.*, 1968, **7**, 288.
19. G. ¨L. Closs, *Advances in Alicyclic Chemistry*, 1966, **1**, 62.
20. K. L. Erickson and J. Wolinsky, *J. Amer. Chem. Soc.*, 1965, **87**, 1142.
21. V. J. Shiner and J. S. Humphrey, *J. Amer. Chem. Soc.*, 1967, **89**, 622.
22. T. L. Gilchrist and C. W. Rees, *J. Chem. Soc. (C)*, 1968, 779.
23. V. Franzen, *Chem. Ber.*, 1962, **95**, 1964.
24. H. E. Simmons, E. P. Blanchard, and R. D. Smith, *J. Amer. Chem. Soc.*, 1964, **86**, 1347.
25. D. Y. Curtin, J. A. Kampmeier, and B. R. O'Connor, *J. Amer. Chem. Soc.*, 1965, **87**, 863.
26. S. D. McGregor and W. M. Jones, *J. Amer. Chem. Soc.*, 1968, **90**, 123. See also R. A. Olofson, S. W. Walinsky, J. P. Marino, and J. L. Jernow, *ibid.*, 1968, **90**, 6554.
27. H. A. Staab and J. Ipaktschi, *Chem. Ber.*, 1968, **101**, 1457.
28. P. T. Keough and M. Grayson, *J. Org. Chem.*, 1962, **27**, 1817.
29. F. Ramirez, H. Yamanaka, and O. H. Basedow, *J. Amer. Chem. Soc.*, 1961, **83**, 173.
30. E. J. Corey, F. A. Carey, and R. A. E. Winter, *J. Amer. Chem. Soc.*, 1965, **87**, 934.

Problems

2–1 Suggest mechanisms for the following:

(a) $CHCl_3$ + $Et_4N^+Br^-$ +

(F. Nerdel and J. Buddrus, *Tetrahedron Lett.*, 1965, 3585).

(b) $\xrightarrow{350°}$ + MeCOOMe

(R. W. Hoffmann and J. Schneider, *Tetrahedron Lett.*, 1967, 4347).

(c)

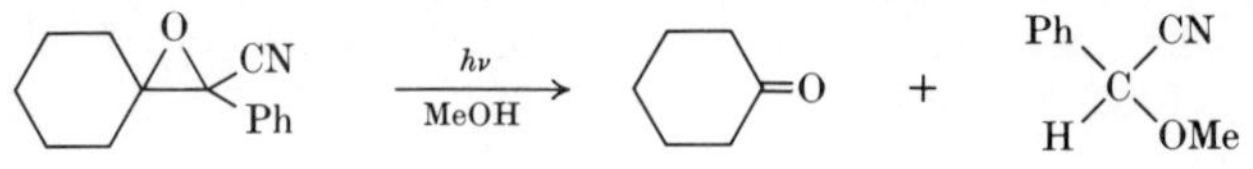

(ref. 9).

(d)

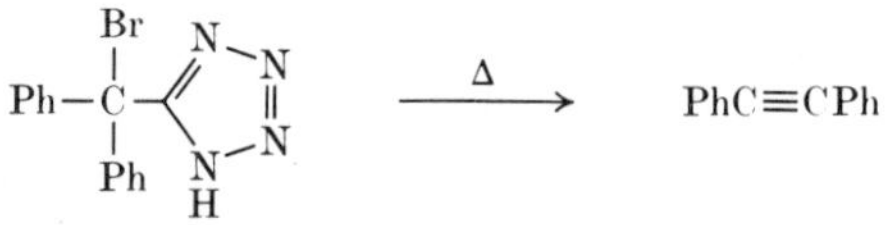

(H. Behringer and M. Matner, *Tetrahedron Lett.*, 1966, 1663).

(e)

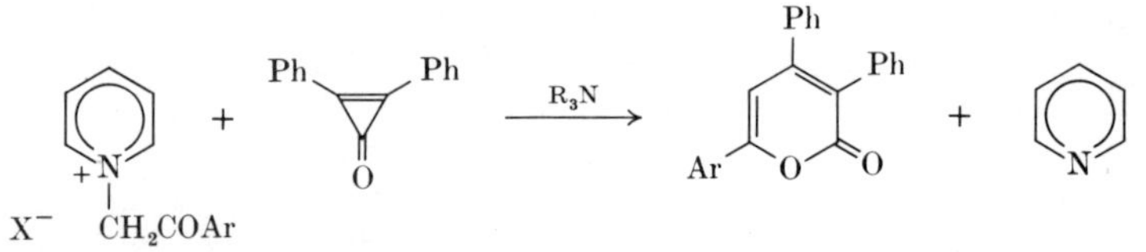

(Th. Eicher and A. Hansen, *Tetrahedron Lett.*, 1967, 1169).

(f)

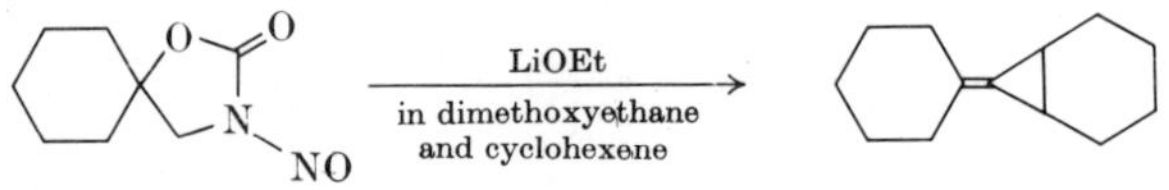

(M. S. Newman and A. O. M. Okorodudu, *J. Amer. Chem. Soc.*, 1968, **90**, 4189).

2–2   When the betaine A is warmed in aprotic solvents (40–80°) in the presence of various electrophiles of the form $X—N_2^+$, the azo compounds B are formed. The rate of evolution of CO_2 is not affected by the nature of the electrophile. The corresponding 4-carboxylate requires a higher temperature for decarboxylation and does not form azo complexes analogous to B. Suggest an explanation.

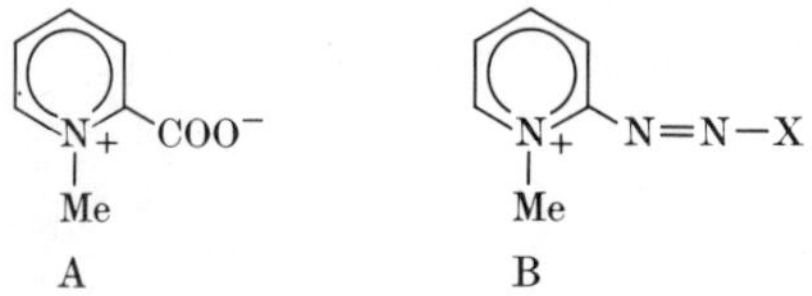

(H. Quast and E. Frankenfeld, *Angew. Chem. Internat. Edn.*, 1965, **4**, 691).

2–3   When Corey's olefin synthesis (Section 2–6) was applied to catechol thionocarbonate (A) in the hope of generating benzyne, no products derived from benzyne were obtained, but, instead, a compound X, $C_{21}H_{12}O_6$, m.p. 178°, which had a strong sharp band in its infrared spectrum at 1756 cm^{-1}. When X was hydrogenated this band disappeared. The nuclear-magnetic-resonance spectrum of the hydrogenation product showed two doublets (τ 3·57 and 5·52, $J = 3$·6 c/sec.), aromatic protons as two singlets (τ 3·07 and 3·13) and a multiplet centred at τ 3·33, the areas of these signals being in the proportions 1:1:4:4:4. With tetracyanoethylene, X gave a 1:1 adduct that showed no infrared band at 1756 cm^{-1}. Acid hydrolysis of X and treatment of the acidic product with diazomethane gave, among other products, catechol, catechol monomethyl ether and compound B.

Suggest a structure for X and a possible mechanism for its formation.

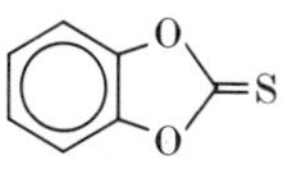

A 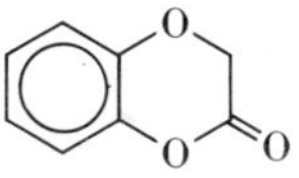   B

(R. Hull and R. Farrand, *Chem. Comm.*, 1965, 164).

2–4 The formation of dichlorocyclopropanes from olefins and phenyl (trichloromethyl) mercury is slow at 80°. The reaction is speeded up enormously by the addition of one mole of sodium iodide, however. The relative reactivities of different olefins are not changed by the addition of sodium iodide. Suggest an explanation for these observations.

(D. Seyferth, M. E. Gordon, J. Y.-P. Mui, and J. M. Burlitch, *J. Amer. Chem. Soc.*, 1967, **89**, 959).

2–5 Diethyl dichloromalonate, $Cl_2C(COOEt)_2$, and potassium t-butoxide in cyclohexene gave the following products: dichloronorcarane (21%), diethyl carbonate (24%), t-butyl ethyl carbonate (35%), and diethyl chloromalonate (5%). There was no di-t-butyl carbonate.

Write a mechanism for the reaction which accounts for these products.
(H. Reimlinger and C. H. Moussebois, *Chem. and Ind.*, 1968, 883).

3 Generation of nitrenes

3–1 Introduction

The chemistry of nitrenes has not been as fully investigated as that of carbenes. Until recently, the only reaction in which nitrenes could generally be postulated as intermediates was the decomposition of azides, a route that parallels the formation of carbenes from diazoalkanes. For most of the other reactions used to generate carbenes, it is possible to find parallel reactions that might involve nitrene intermediates. These have usually been much less thoroughly explored, however, and alternative mechanisms, not involving nitrenes, can often be written for them. There are also a few reactions, such as the reduction of nitro and nitroso compounds and the oxidation of amines, that may involve nitrenes and that have no direct parallels in carbene chemistry.

3–2 From azides

Azides decompose on heating, usually at temperatures between 100° and 200°, or on irradiation. Nitrogen is lost and the reaction products can often be rationalized by postulating nitrene intermediates. The most common reactions are abstractions of hydrogen from the solvent, insertions, and intramolecular rearrangements. With olefins, aziridines may be formed, but nitrenes are probably not intermediates, at least with polar olefins. The rate of thermolysis of phenyl azide in reactive olefins is much faster than in an inert solvent such as tetralin or nitrobenzene, showing that the olefin is involved in the rate-determining step. The azide adds to the double bond to give a triazoline, which can often be isolated in mild conditions; on heating, the triazoline may lose nitrogen to give an aziridine:

$$PhN_3 + H_2C=C\underset{COOMe}{\overset{Me}{<}} \xrightarrow{\Delta} \underset{Ph}{\overset{Me}{\underset{}{N}}}\underset{}{N} \longrightarrow products$$

$$PhN_3 + \bigcirc \xrightarrow{\Delta} \bigcirc N \longrightarrow \bigcirc N-Ph$$

In solvents such as tetralin, or in the gas phase, simple azides almost certainly decompose to give nitrene intermediates on thermolysis or photolysis. The parent nitrene, $:\ddot{N}H$, has been detected spectroscopically in the photolysis of hydrazoic acid, HN_3, both in the gas phase and in matrices at very low temperatures. Similarly, direct spectroscopic evidence for methylnitrene has been obtained in the gas-phase photolysis of methyl azide. Aryl azides decompose in solution at a rate that is independent of the solvent (with the exception of the olefinic solvents mentioned above), and the effect of substituents in the benzene ring on the rate of decomposition is small.

Halogenonitrenes and cyanonitrene can also be generated from the corresponding azides. Cyanogen azide is unusual in that it decomposes at a temperature of between 40° and 50°, much lower than for most other azides. This is presumably because of the special symmetry and stability of the nitrene :[1]

$$NCN_3 \longrightarrow \cdot N{=}C{=}N\cdot + N_2$$

For many other azides, however, the evidence that nitrenes are intermediates in their decomposition is not convincing. Azides with a nucleophilic centre suitably placed within the molecule lose nitrogen on photolysis or thermolysis to give products that can be rationalized either by a nitrene mechanism or by a concerted cyclization and loss of nitrogen. When five-membered rings are formed, the latter seems more likely. Some examples are shown :

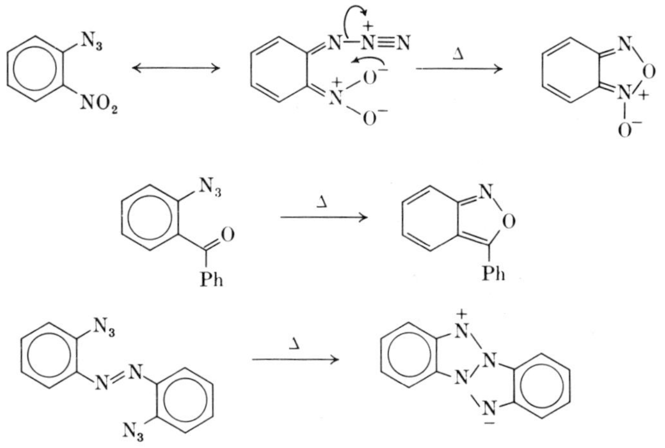

For vinyl azides, either mechanism is possible. α-Styryl azide gives 2-phenylazirine, but β-styryl azide gives phenylacetonitrile, which may be formed through the isomeric 3-phenylazirine :[2]

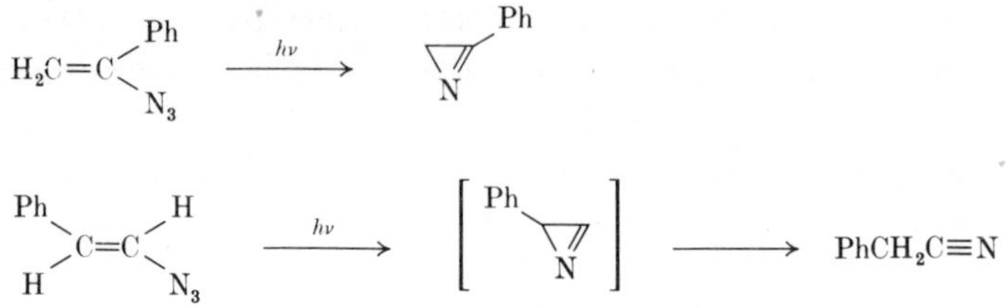

Acylnitrenes are particularly important, as they have been proposed as intermediates in the Curtius rearrangement of carbonyl azides to isocyanates:

$$\text{RCON}_3 \xrightarrow{\Delta \text{ or } h\nu} \text{R—N=C=O} + \text{N}_2$$

There is evidence that acylnitrenes can be intermediates in the photolysis of these azides. The nitrenes can be trapped with olefins or with saturated hydrocarbons.

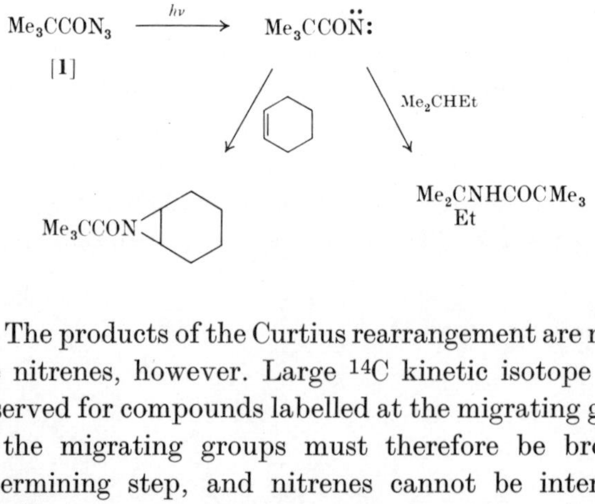

The products of the Curtius rearrangement are not formed through the nitrenes, however. Large ^{14}C kinetic isotope effects have been observed for compounds labelled at the migrating groups.[3] The bonds to the migrating groups must therefore be broken in the rate-determining step, and nitrenes cannot be intermediates. Similar isotope effects were found in the related Hofmann and Lossen rearrangements. Other evidence against a nitrene mechanism for the Curtius rearrangement has come from Lwowski, who has shown that, in standard photolysis conditions, the yield of isocyanate from pivaloyl azide [1] is independent of the nature of the solvent.[4] If the isocyanate were formed from the nitrene, its yield would vary with the ability of the solvent to trap the nitrene. There must therefore be two independent modes of reaction of the azides in photolyses: loss of nitrogen to give the nitrene, which can be trapped but which does not rearrange (path A), and concerted migration and loss of nitrogen to give the isocyanate (path B):

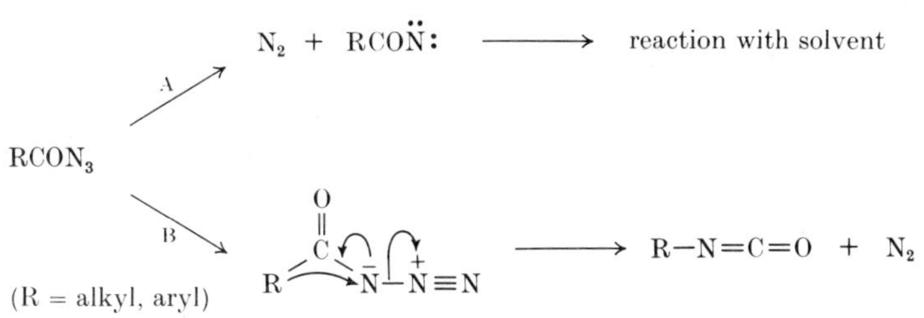

$$N_2 + RCO\ddot{N}: \longrightarrow \text{reaction with solvent}$$

$$RCON_3$$

(R = alkyl, aryl)

$$R-N=C=O + N_2$$

No nitrene products have yet been found from *thermolyses* of acyl azides, so these reactions must go entirely by path B, to give rearranged products by a concerted mechanism.

Alkoxycarbonylnitrenes, $:\ddot{N}COOR$, are produced from the corresponding azides both by photolysis and by thermolysis, however. The same products of insertion or addition are formed in approximately the same ratios, either by photolysis or by thermolysis; Curtius rearrangement is only a minor reaction.[5] Sulphonyl azides, RSO_2N_3, also probably decompose thermally to nitrenes, since their thermolysis, which starts at 120–140°, is usually independent of the solvent used. The products, which are mainly those of hydrogen abstraction or of insertion into the solvent, are best explained by a nitrene mechanism.

A few bis-azides are known. 1,2-Diazidobenzene and its derivatives decompose to 1,4-dicyanobutadienes, and the geminal diazide $Ph_2C(N_3)_2$ gives diphenylcarbene on photolysis:[6]

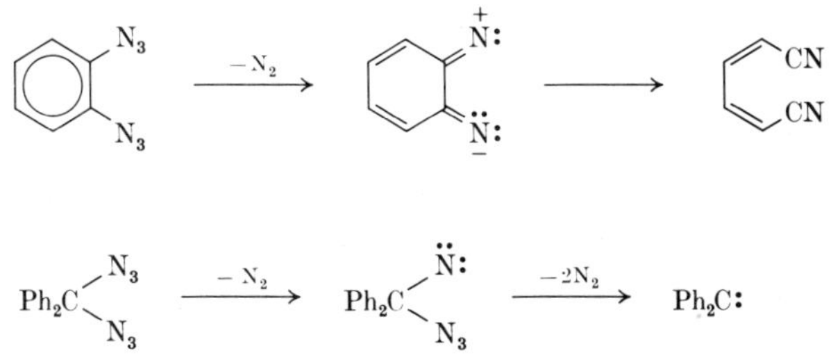

In both, the loss of nitrogen is probably stepwise; the intermediate mononitrene has been detected in the latter reaction by electron spin resonance.

Just as with the diazoalkane decompositions, the thermolyses of azides can by catalysed by copper and by derivatives of other transition metals. These reactions probably involve complexes of the azides and of the nitrenes with the transition metals (Chapter 7).

3–3   Other thermolytic and photolytic routes

Ketens, epoxides, and several similar compounds can give carbenes on photolysis (Chapter 2). There are analogous nitrogen compounds that can be used as sources of nitrenes. For example, isocyanates can be photolysed to give nitrenes and carbon monoxide.[2] This is analogous to the keten route to carbenes.

$$RN{=}C{=}O \xrightarrow{\;h\nu\;} R\ddot{N}{:} + CO$$

Oxaziridines, such as [2], have been photolysed to give nitrenes, which can be detected by electron spin resonance at low temperatures.[7] This interesting reaction parallels that in which carbenes are formed from epoxides. Other reactions are known in which nitrenes can be formed by the photolysis of N-oxides, such as the quinonedianil NN'-dioxide [3][8] and the nitrile oxide [4].[9] These reactions probably go by a similar mechanism, through cyclic oxaziridine and oxazirine intermediates, respectively.

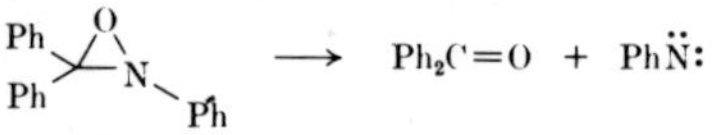

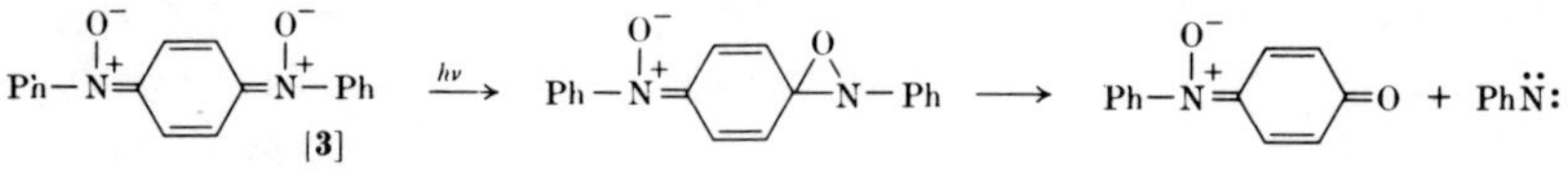

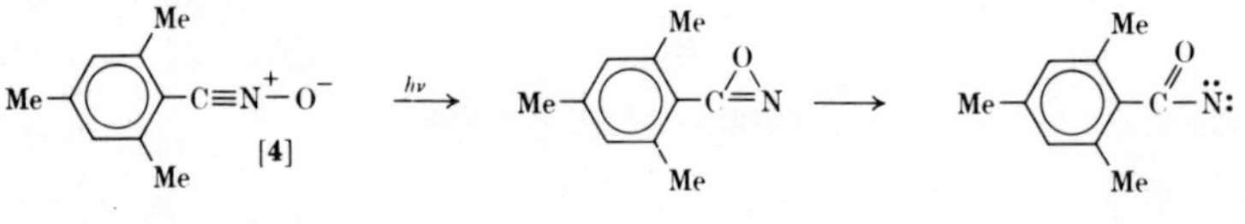

Acylnitrenes have also been formed from the thermal or photochemical fragmentation of the dioxazole derivatives [5]. In dimethyl sulphoxide, thermal decomposition gives products derived from the unrearranged and the rearranged nitrenes, but photolysis in dimethyl sulphide gives products derived only from the unrearranged nitrenes.[10] Whether the rearranged products come from the nitrenes, or from a concerted decomposition analogous to that of the azides, remains open.

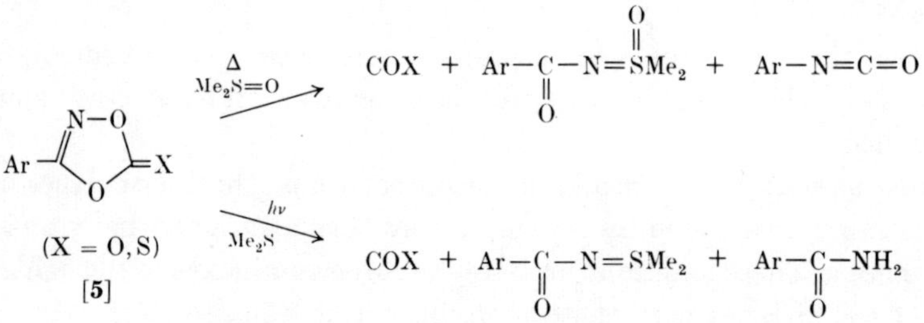

Phosphinimines can also give nitrenes on photolysis:[11]

$$Ph_3P=N-CMe_3 \longrightarrow Ph_3P + :\ddot{N}-CMe_3$$

This reaction is analogous to the formation of carbenes from ylids.

3–4 From nitrogen anions

α-Elimination reactions, usually base-induced, are an important source of carbenes. Analogous reactions can be envisaged for nitrenes:

$$R-N\begin{matrix} H \\ \\ X \end{matrix} \xrightarrow[\text{B:}]{\text{base}} R-\ddot{N}: + X^- + BH^+$$

Reactions of this type are less common in nitrogen chemistry than in carbon chemistry, and the mechanisms are usually in doubt. It is likely that the first step is the removal of a proton to form a nitrogen anion. The slow step may then be the formation of a nitrene and an anion X^-, or there may be a concerted reaction involving both the substituent R and the leaving group X. The products of these reactions can usually be rationalized by either mechanism. Where no positive evidence is available, we can only guess whether nitrenes are likely to be intermediates or not by comparing the products with those obtained from reactions where the mechanism is known.

The Hofmann and Lossen rearrangements are the best-known examples of reactions of this type. In the Hofmann rearrangement, an N-bromoamide or N-chloroamide is made to react with a base to give a primary amine via an intermediate isocyanate:

$$R-\overset{O}{\overset{\|}{C}}-NHBr \xrightarrow{-OH} R-N=C=O \longrightarrow RNH_2$$

The Lossen rearrangement is similar; the reaction is between a hydroxamic acid ester and a strong base:

$$R-\overset{O}{\overset{\|}{C}}-NHOCOR \xrightarrow{-OH} R-N=C=O \longrightarrow RNH_2$$

The mechanism of both is a concerted rearrangement of the nitrogen anion, with simultaneous migration of the alkyl group to nitrogen and loss of the leaving group, just as for the related Curtius rearrangement. This is deduced from the fact that nitrenes have never been trapped from these reactions, and, more directly, from kinetic isotope effects, which show that the group R participates in the rate-determining step:[3]

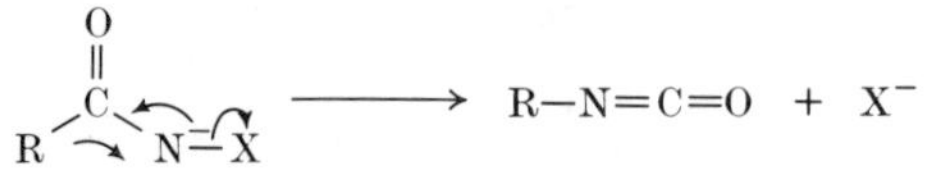

$$R-N=C=O \; + \; X^-$$

There are several other reactions of this type for which mechanisms involving nitrenes are possible, but unlikely. The reaction of chloramine with bases is one; two possible mechanisms are shown:

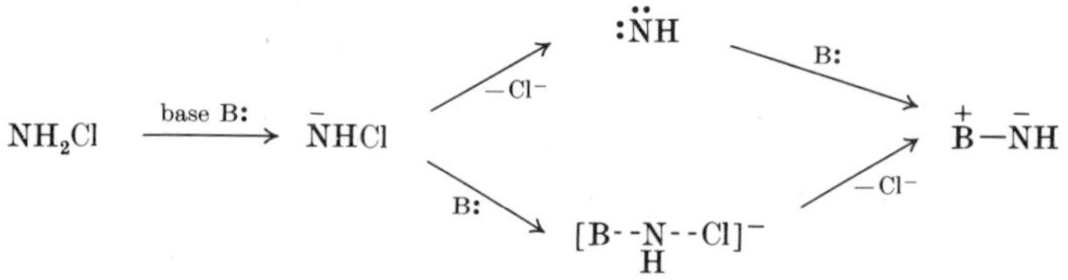

An example of a reaction of this kind is the Raschig synthesis of hydrazine from ammonia and hypochlorite, in which the primary process is the formation of chloramine:

$$NH_2Cl + NH_3 \xrightarrow{\;^-OH\;} H_2NNH_2 + Cl^- + H_2O$$

One piece of evidence against the intermediacy of nitrenes from chloramines and bases is that photolysis of the corresponding azides gives different products:

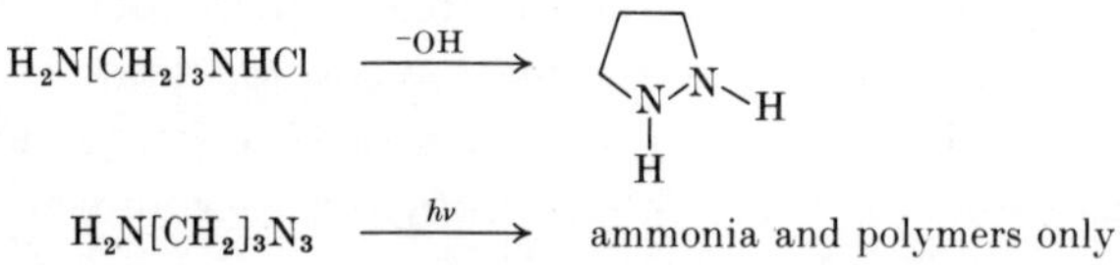

$$H_2N[CH_2]_3NHCl \xrightarrow{\;^-OH\;}$$

$$H_2N[CH_2]_3N_3 \xrightarrow{\;h\nu\;} \text{ammonia and polymers only}$$

If the same intermediates were involved, the same products might be expected.

Similarly, the reactions of hydroxylamine-O-sulphonic acid, H_2NOSO_3H, with bases are not likely to involve nitrene as an intermediate; the products are not those that one would expect from a free nitrene. Cyclohexanone gives an oxaziridine, for example:

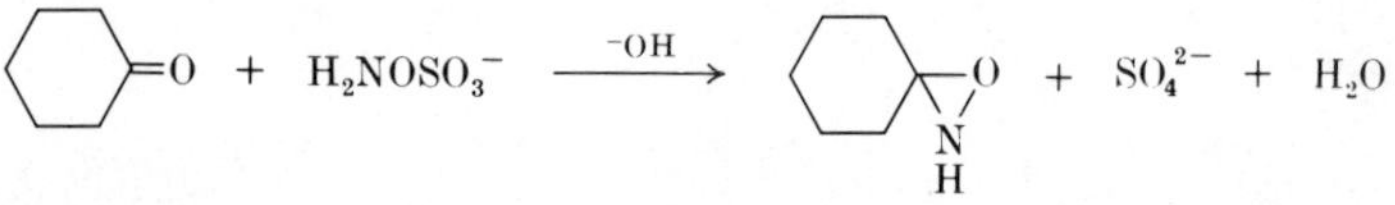

$$\rangle=O \; + \; H_2NOSO_3^- \xrightarrow{\;^-OH\;} \quad + \; SO_4^{2-} \; + \; H_2O$$

Some base-catalysed α-eliminations probably do give nitrenes, however. The best-established example is the generation of ethoxycarbonylnitrene from N-(p-nitrobenzenesulphonyloxy)urethane [6]. The product distribution is very similar to that from the photolysis or thermolysis of ethoxycarbonyl azide, and the nitrene is the most likely common intermediate.

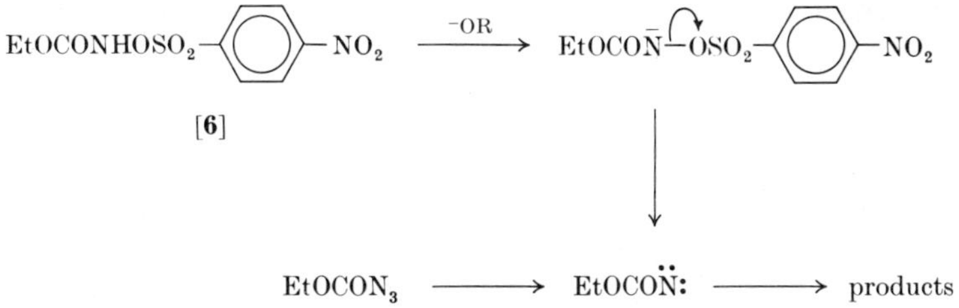

$$EtOCONHOSO_2 \!-\!\!\langle \bigcirc \rangle\!-\! NO_2 \quad \xrightarrow{-OR} \quad EtOCO\bar{N}\!-\!OSO_2\!-\!\langle \bigcirc \rangle\!-\! NO_2$$

[6]

$$EtOCON_3 \quad \longrightarrow \quad EtOCO\ddot{N}: \quad \longrightarrow \quad products$$

Aminonitrenes are probably generated in the thermolysis or photolysis of salts of 1,1-disubstituted toluene-p-sulphonylhydrazides, a reaction related to the formation of diazoalkanes and carbenes from salts of toluene-p-sulphonylhydrazones.

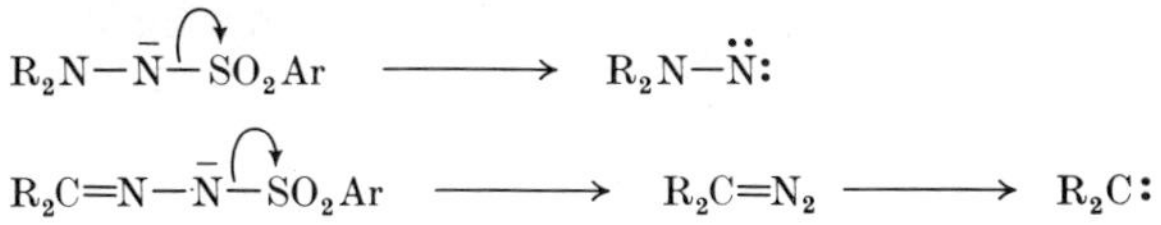

$$R_2N\!-\!\bar{N}\!-\!SO_2Ar \quad \longrightarrow \quad R_2N\!-\!\ddot{N}:$$

$$R_2C\!=\!N\!-\!\bar{N}\!-\!SO_2Ar \quad \longrightarrow \quad R_2C\!=\!N_2 \quad \longrightarrow \quad R_2C:$$

This reaction has been carried out with a wide range of cyclic and acyclic hydrazides; the products are determined by the substituents on the hydrazide, but they can be explained by a nitrene mechanism. When comparisons with other methods of generating the same nitrenes have been made, the products are usually the same. In the examples below, identical products were obtained by oxidation of the corresponding hydrazines:

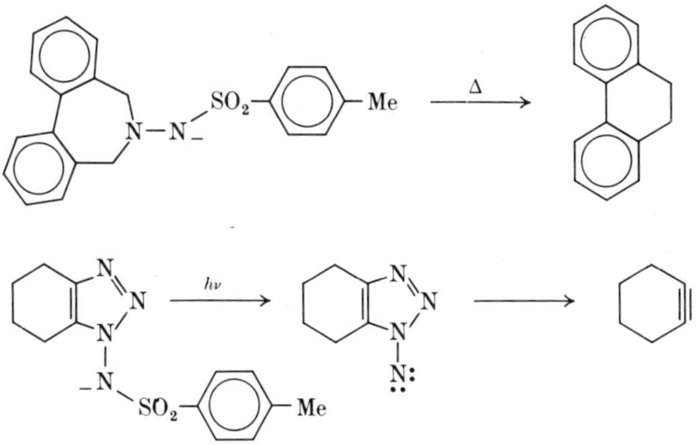

3–5 Oxidative routes

Nitrenes are possible intermediates in the oxidation of compounds containing NH_2 groups:

$$R\!-\!NH_2 + [O] \longrightarrow R\!-\!\ddot{N}: + H_2O$$

There are many other possible mechanisms, some very complex, and in general oxidation is neither a practicable nor a very likely method of generating nitrenes. With a few systems, however, the products are best rationalized by assuming that nitrenes are generated.

Even in these systems other mechanisms are possible, and the evidence for nitrenes is not strong.

1,1-*Disubstituted hydrazines.* The oxidation of 1,1-disubstituted hydrazines can give a great variety of products. The nature of the products is sensitive to quite minor changes in the structure of the substrate, type of oxidant, and reaction conditions. Despite these variations, most of the products can be rationalized by postulating aminonitrenes as intermediates:

$$R_2N\text{—}NH_2 \xrightarrow{\text{[O]}} R_2N\text{—}\ddot{N}: \longleftrightarrow R_2\overset{+}{N}\!\!=\!\!\overset{-}{N}$$

The oxidant that has been used most extensively is mercuric oxide. The normal products of these oxidations with mercuric oxide are tetrazenes, though many others have also been found:

$$2R_2N\text{—}NH_2 \xrightarrow{\text{HgO}} R_2N\text{—}N\!\!=\!\!N\text{—}NR_2$$

A reasonable mechanism for formation of the tetrazene is an attack of the aminonitrene on unchanged hydrazine.

Other oxidants that have been used include activated manganese dioxide, nickel peroxide, bromine, *N*-bromo- and *N*-chloro-succinimide, and potassium permanganate, but the best one is lead tetra-acetate, which oxidizes the hydrazines rapidly and cleanly. With NH_2 groups, lead tetra-acetate probably first forms a complex, $RNHPb(OAc)_3$, which may then fragment to the nitrene, lead di-acetate, and acetic acid, or which may itself be the reactive intermediate. If the oxidation of the hydrazines with lead tetra-acetate is carried out in the presence of nucleophilic or electrophilic olefins, aziridines may be formed, often in high yield:[12]

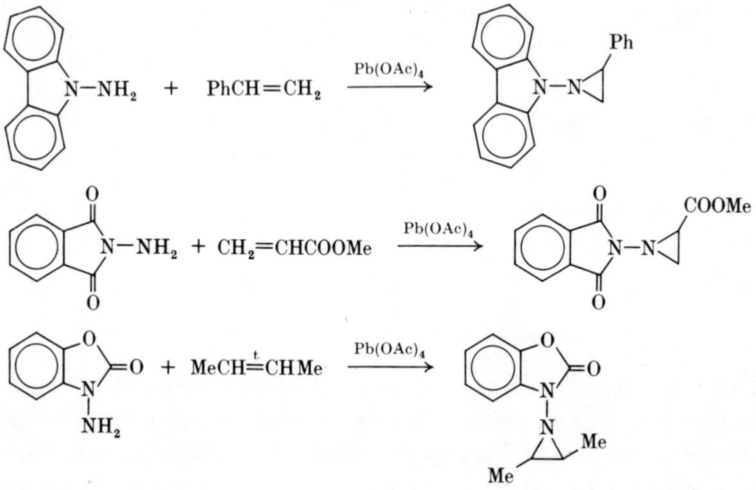

The aziridines are formed stereospecifically. Not all N-amino compounds can form aziridines, however; intramolecular rearrangements occur preferentially in 1,1-dialkylhydrazines, for example.

Primary amines. There is no compelling evidence that nitrenes are intermediates in the oxidation of simple primary amines, though some products, such as azo compounds, can be rationalized on the basis of a nitrene mechanism:

$$ArNH_2 \longrightarrow Ar\ddot{\underset{\cdot\cdot}{N}}: + ArNH_2 \longrightarrow ArNHNHAr \longrightarrow ArN{=}NAr$$

With olefins, aziridines cannot normally be formed from primary amines, unless the addition is intramolecular, as, for example, with 4-(β-aminoethyl)cyclohexene [7]:[13]

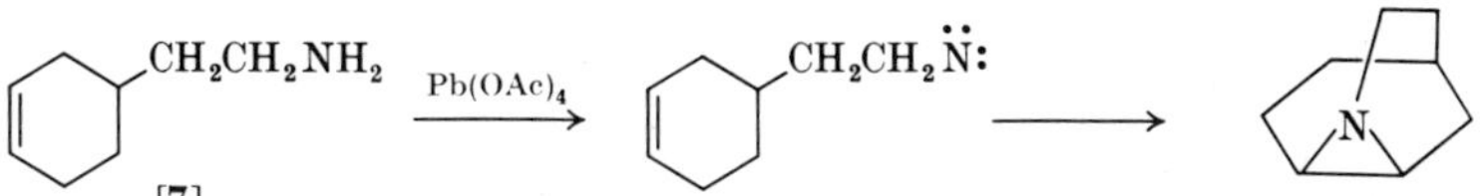

It is possible that nitrenes are intermediates in this reaction.

Primary amides. Although amides can be oxidized by lead tetraacetate, there is no evidence that nitrenes are intermediates; they cannot be trapped with olefins, for example. When rearrangements of the Hofmann type occur, they probably go by a concerted mechanism.[14]

3–6 Reductive routes

When heated with certain reducing agents, aromatic nitro and nitroso compounds give products that may be formed via nitrene intermediates. The synthesis of carbazole from 2-nitrobiphenyl and ferrous oxalate or triethyl phosphite is an example:

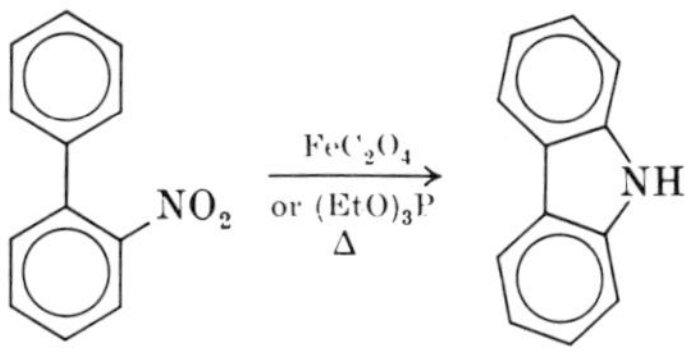

The best deoxygenating agents are phosphines and phosphites. It is by no means certain that all the reactions of this type go via nitrenes; the best evidence that some do is a comparison of the products obtained from the deoxygenation of nitro compounds with phosphites and the photolysis of the corresponding azides.[15] From

o-n-butylnitrobenzene and o-n-butylazidobenzene, for example, the product distributions are very similar:

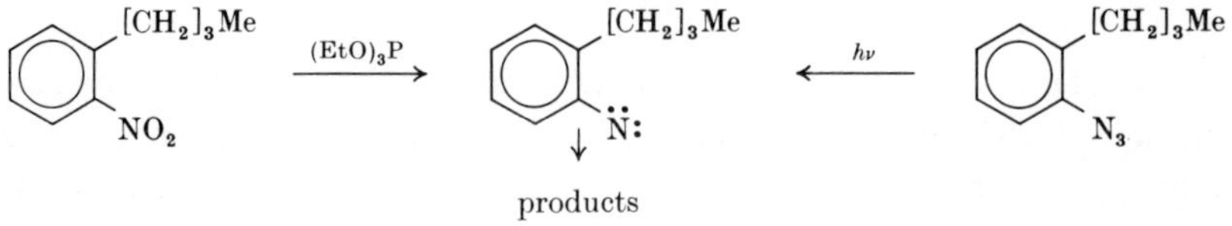

products

The products include those of insertion into the side chain. In particular, if the starting material has an optically active carbon in the side chain, then the optical activity is retained in the cyclized product (both from the azide and the nitro compound), indicating that the product is most probably formed from a singlet nitrene by insertion:

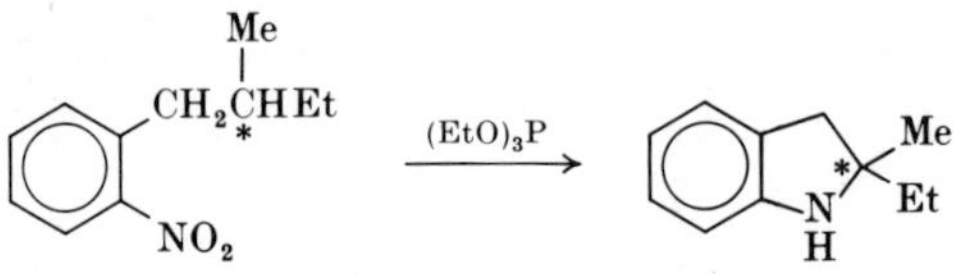

If ferrous oxalate is used as the reducing agent, no optically active product is obtained. This throws doubt on the intermediacy of nitrenes in reactions involving the use of ferrous oxalate.

The triethyl phosphite reduction of nitro compounds has been applied to a variety of systems; some examples are shown below:

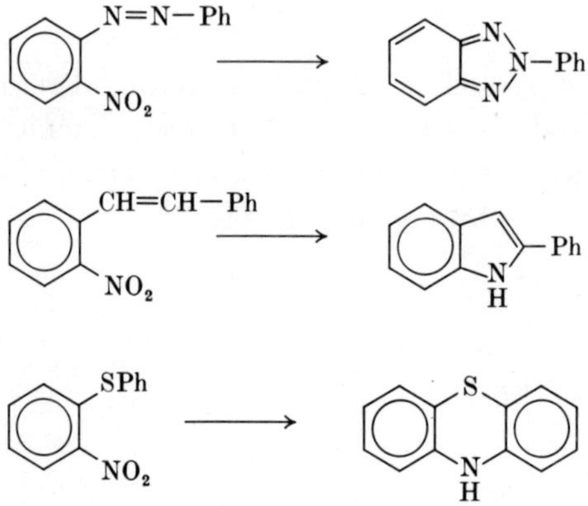

It is a useful synthetic route to nitrogen heterocyclic compounds.

Nitroso compounds react very much more readily with triethyl phosphite, and nitrenes are probably intermediates in the reactions.

Triethyl phosphite, triphenylphosphine, and tributylphosphine have been used to deoxygenate nitrosobenzene. In amine solvents,

the products are similar to those formed by photolysis of the corresponding azides, namely N-alkyl derivatives of 2-amino-$3H$-azepines:[16]

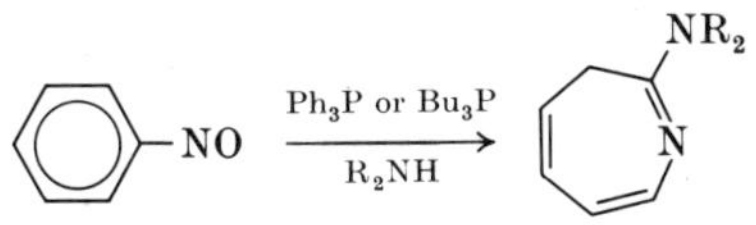

In principle, the deoxygenation of N-nitroso compounds should be
another way of making aminonitrenes. Not much is known about
this reaction: apparently triphenylphosphine does not deoxygenate
N-nitroso compounds, but sodium dithionite and lithium in liquid
ammonia have been used:

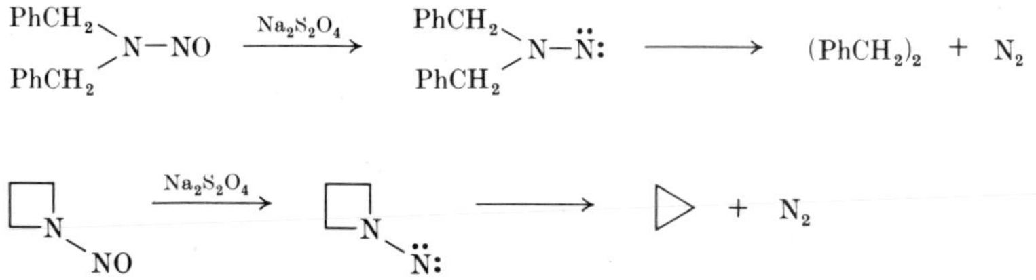

The evidence for aminonitrenes in these reactions is again simply
based on the nature of the products.

3–7 Other routes

An interesting reaction in which aminonitrenes are apparently
formed directly from secondary amines takes place between the
amines and Angeli's salt, sodium nitrohydroxylamate, $Na_2N_2O_3$, in
acid.[17] The products obtained are those expected from the nitrenes:
dibenzylamine, for example, gives dibenzyl, and piperidine gives
some tetrazene:

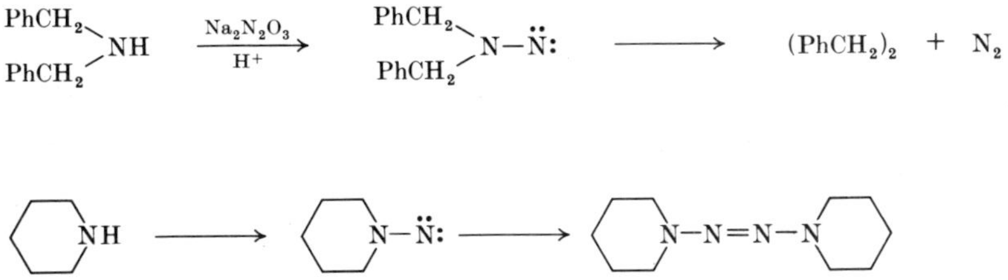

(and other products)

Difluoramine, HNF_2, is another reagent that can convert secondary
amines into aminonitrenes. Again, dibenzylamine gives dibenzyl.

Aziridine is converted into ethylene, and azetidine into cyclopropane, also presumably through the nitrenes:

With substituted aziridines, the olefins are formed stereospecifically.

References

1. A. G. Anastassiou and H. E. Simmons, *J. Amer. Chem. Soc.*, 1967, **89**, 3177.
2. J. H. Boyer, W. E. Krueger, and G. J. Mikol, *J. Amer. Chem. Soc.*, 1967, **89**, 5504.
3. *Chem. and Eng. News*, 1968, Jan. 1st., 28.
4. W. Lwowski, *Angew. Chem. Internat. Edn.*, 1967, **6**, 897.
5. R. Puttner, W. Kaiser, and K. Hafner, *Tetrahedron Lett.*, 1968, 4315.
6. J. H. Hall and E. Patterson, *J. Amer. Chem. Soc.*, 1967, **89**, 5856.
   L. Barash, E. Wasserman, and W. A. Yager, *J. Amer. Chem. Soc.*, 1967, **89**, 3931.
7. J. S. Splitter and M. Calvin, *Tetrahedron Lett.*, 1968, 1445.
8. C. J. Pedersen, *J. Amer. Chem. Soc.*, 1957, **79**, 5014.
9. G. Just and W. Zehetner, *Tetrahedron Lett.*, 1967, 3389.
10. J. Sauer and K. K. Mayer, *Tetrahedron Lett.*, 1968, 319.
11. H. Zimmer and M. Jayawant, *Tetrahedron Lett.*, 1966, 5061.
    T. L. Gilchrist, F. J. Graveling, and C. W. Rees, unpublished work.
12. R. S. Atkinson and C. W. Rees, *Chem. Comm.*, 1967, 1230.
    D. J. Anderson, T. L. Gilchrist, D. C. Horwell, and C. W. Rees, *Chem. Comm.*, 1969, 146.
13. W. Nagata, S. Hirai, K. Kawata, and T. Aoki, *J. Amer. Chem. Soc.*, 1967, **89**, 5045.
14. B. Acott, A. L. J. Beckwith, and A. Hassanali, *Austral. J. Chem.*, 1968, **21**, 185.
15. J. I. G. Cadogan, *Quart. Rev.*, 1968, **22**, 222.
16. R. A. Odum and M. Brenner, *J. Amer. Chem. Soc.*, 1966, **88**, 2074.
17. D. M. Lemal and S. D. McGregor, *J. Amer. Chem. Soc.*, 1966, **88**, 1335.

Problems

3–1 Photolysis of phenyl azide in cyclohexene failed to give any aziridine, but thermolysis (80°) gave the aziridine in high yield. Explain.

(W. von E. Doering and R. A. Odum, *Tetrahedron*, 1966, **22**, 81; K. R. Henery-Logan and R. A. Clark, *Tetrahedron Lett.*, 1968, 801).

3–2 Suggest mechanisms for the following:

(a)
O⁻
O⁻
Ph—N
N—Ar
hν
O
O
+
PhN=NPh
+ PhN=NAr + ArN=NAr

(ref. 8).

(b)
Ph
NO2
O
N—O—C
Et3N
benzene
N
H
NO2
(minor product)

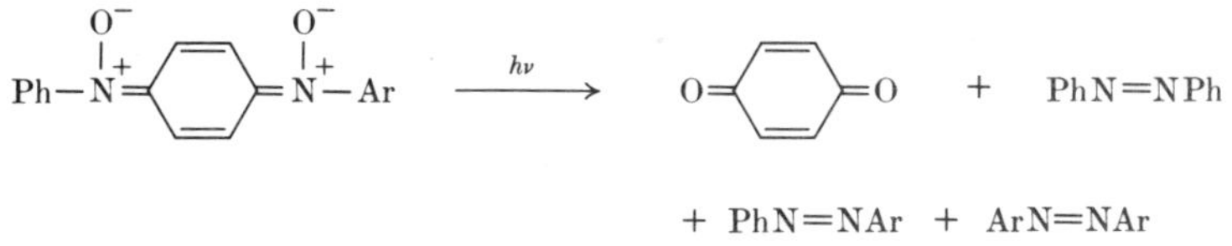

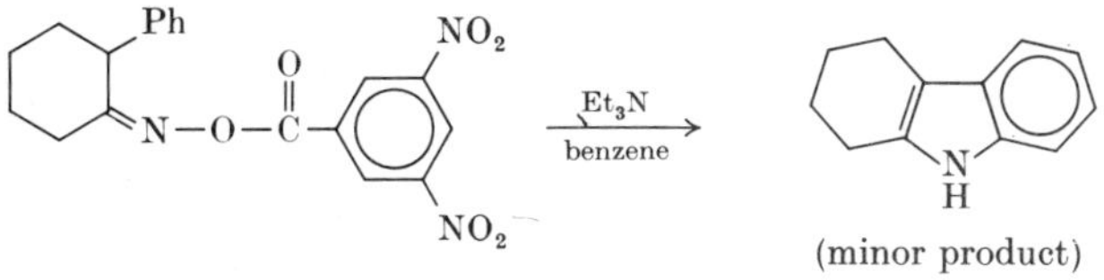

(E. Wenkert and B. F. Barnett, *J. Amer. Chem. Soc.*, 1960, **82,** 4671).

3–3 Treatment of 1,1-disubstituted sulphamides $R_2NSO_2NH_2$ with aqueous alkaline hypochlorite gave 1,1-disubstituted hydrazines R_2NNH_2. Suggest possible mechanisms and indicate how they might be differentiated.

(R. Ohme and H. Preuschhof, *Annalen*, 1968, **713,** 74).

4 Generation of arynes

4-1 Introduction

Arynes are formed by the removal of two adjacent substituents from an aromatic nucleus. If the two substituents are lost in a stepwise reaction, then aryl radicals, cations or anions will be intermediates. Of these, aryl anions are the most important in practice. Another useful general route to arynes is the fragmentation of heterocyclic rings fused to the *ortho* positions of the aromatic system; these eliminations are probably concerted processes.

4-2 From aryl anions

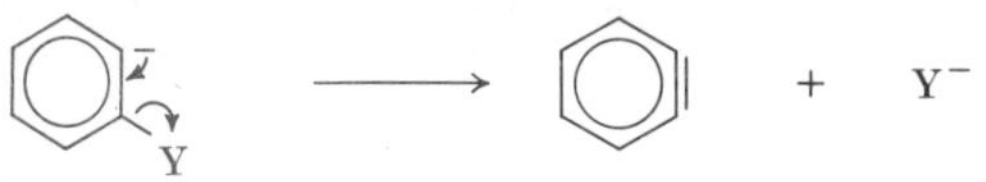

Since 1902 there have been suggestions that arynes might be intermediates in various aromatic substitutions. Wittig was the first to put the suggestion clearly and convincingly, during the 1940s. He studied the reaction of halogenobenzenes with phenyl-lithium and found that biphenyl was formed, but that its rate of formation from fluorobenzene was greater than from the other halogenobenzenes. A straight displacement of fluoride by the phenyl anion was therefore very unlikely. Wittig suggested that the inductive effect of the fluorine facilitated removal of the *o*-hydrogen and its replacement by lithium. Thus, an *o*-fluorophenyl anion was generated, which then lost fluoride to give benzyne. Phenyl-lithium could then react rapidly with the benzyne to give *o*-lithiobiphenyl, which Wittig showed to be the primary product of the reaction:

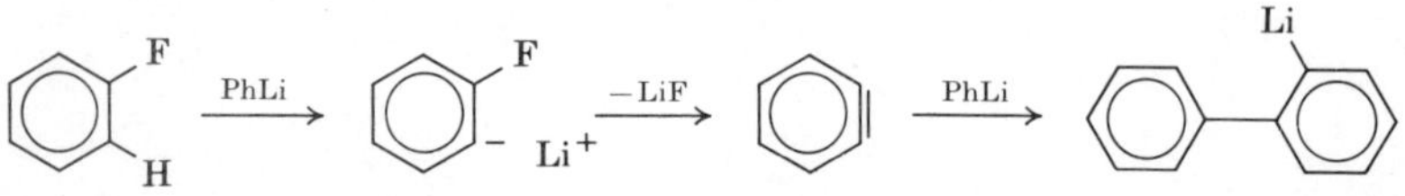

In this reaction, phenyl-lithium has a dual role: it acts as a strong base to remove the *o*-hydrogen, generating the *o*-fluorophenyl anion, and it acts as a nucleophile in its addition to benzyne. J. D. Roberts established the presence of arynes in reactions of this type during some classic experiments with halogenobenzenes and potass-amide in liquid ammonia. Here the product is aniline, formed by nucleophilic attack of NH_3, or of $^-NH_2$, on the intermediate benzyne. Roberts showed that benzyne was an intermediate by using chloro-benzene labelled at C-1 with ^{14}C. In the aniline formed, about half had the label at C-1 and the rest had the label at C-2, indicating that the two positions had become equivalent during the reaction:

The slow step in the formation of benzyne from the halogeno-benzenes may be either the removal of a proton, or the loss of halide from the o-halogenophenyl anion. Re-protonation of the anion will compete with loss of halide; the relative rates will depend on the reaction conditions, particularly the nature of the solvent, and also on the nature of the halogen. The rates of loss of halides from the anions are in the order Br>Cl>F. With potassamide in liquid ammonia, for example, the rate of loss of bromide is faster than protonation of the anion, the rate of loss of chloride is about the same, and the rate of loss of fluoride is so much lower than that of pro-tonation that no aniline is formed from fluorobenzene.

Since the classic work of Wittig and of Roberts, many similar routes to arynes have been developed with a variety of strong bases and of halogenated aromatic substrates. The strength of the base needed to remove a proton from an aromatic system depends very much on the nature of the aromatic system: whereas very strong bases are needed with the halogenobenzenes, organic bases like piperidine may be sufficiently strong to remove a proton from derivatives of coumarin and similar heterocycles.

Base systems that have been commonly used include lithium alkyls and lithium aryls in ether, lithium amides, sodamide, and potass-amide in ether or in the presence of the free amine, potassium t-butoxide in dimethyl sulphoxide, and potassium hydroxide in di-phenyl ether. The reaction has been used to generate dehydro-aromatic intermediates from derivatives of benzene, naphthalene, phenanthrene, and a variety of heterocyclic systems. It is also a good way of making cycloalkynes. Some examples will illustrate the scope of the reaction:

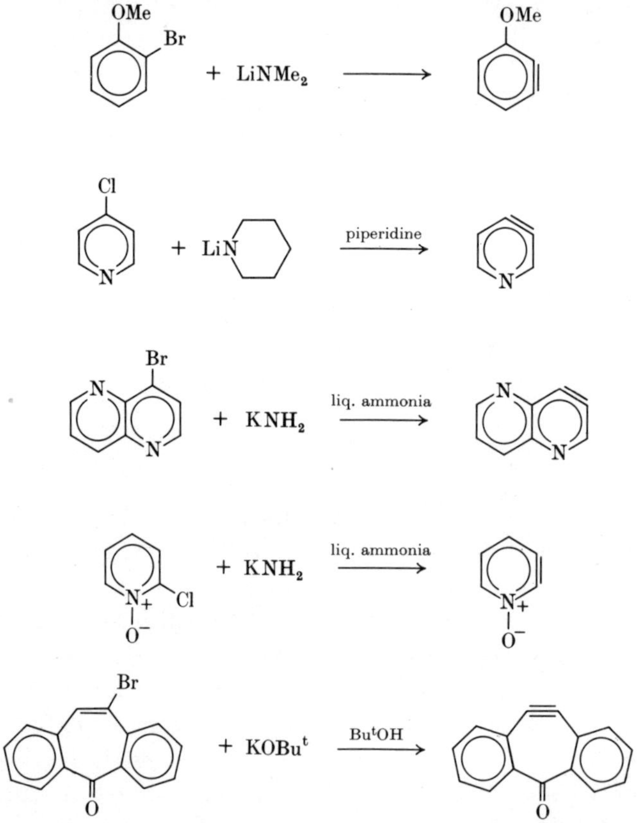

In many systems two different arynes are possible. *meta*-Substituted halogenobenzenes are the simplest examples; others include 2-halogenonaphthalenes and 3-halogenopyridines.

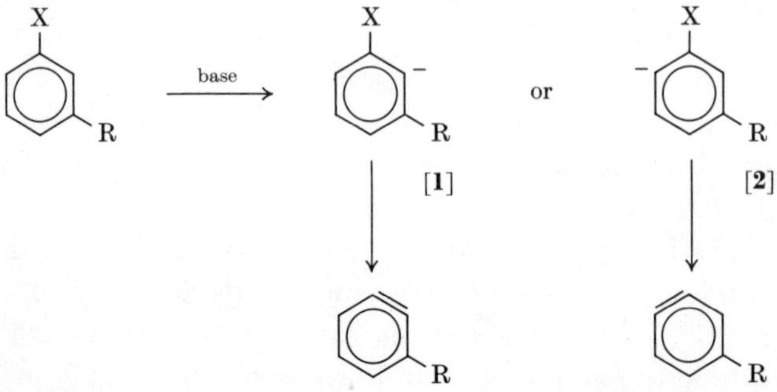

The effect of the substituent R in determining the ratio of the two possible arynes is mainly an inductive one. If R is inductively electron-withdrawing, the anion [1] will be formed in preference to [2]; if R is electron-releasing, [2] will be preferred. The fact that these are the preferred ions for a given substituent does not necessarily mean that the corresponding arynes are also formed preferentially, however. If the loss of halide from the anion is the rate-determining step in the reaction, then it is the relative rates of halide loss from the two possible anions that will determine the ratio of arynes. This is nicely illustrated by the reactions of *m*-bromotoluene and of *m*-chloro-toluene with potassamide. In the former, the rate-determining step is the formation of the anions, but, in the latter, it is the loss of chloride from the anions. Since the methyl group is electron-releasing, the anion [4] is formed preferentially, but the rate of loss of chloride from [4] is slower than from [3]. Hence, *m*-bromotoluene gives mainly the aryne [6], but *m*-chlorotoluene gives mainly [5].

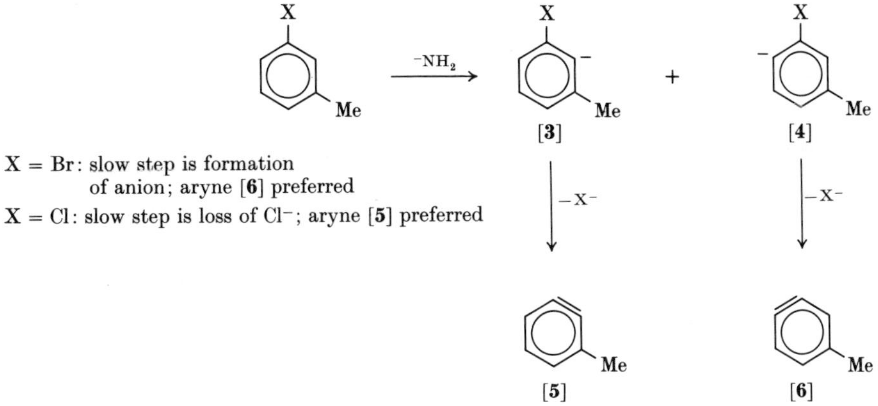

X = Br: slow step is formation
      of anion; aryne [6] preferred
X = Cl: slow step is loss of Cl⁻; aryne [5] preferred

From 3-halogenopyridines, 3,4-dehydropyridine is generated almost exclusively. This is related to the preferential loss of the 4-hydrogen; the 2-hydrogen is exchanged only very slowly for deuterium in basic conditions.[1] Similarly, 2-halogenonaphthalenes give only 1,2-dehydronaphthalene.

Hitherto we have tacitly assumed that the halogenated aromatic compounds always react with the bases by an aryne mechanism, but this is not so. An addition–elimination mechanism is also possible, especially in aromatic systems activated to nucleophilic substitution such as pyridine derivatives:

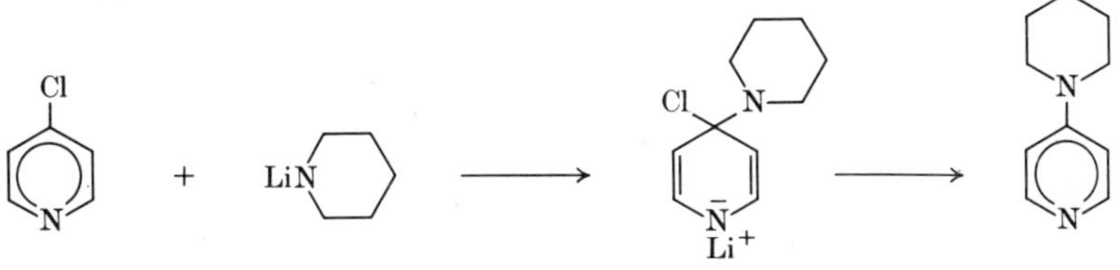

An obvious difference between the aryne mechanism and the above addition–elimination mechanism is that the former is much more likely to lead to *cine*-substitution; that is, substitution at a carbon adjacent to the one carrying the leaving group. The isolation of products of *cine*-substitution is often taken as good evidence for the aryne mechanism. This is not an infallible criterion, however; in particular, *cine*-substitution can occur by other mechanisms in five-membered-ring heterocyclic compounds. For example, 2-bromothiophen is converted into 3-aminothiophen by potassamide in liquid ammonia, but not via 2,3-dehydrothiophen. The anion [7] is rather stable and can react with more 2-bromothiophen to give 2,3-dibromothiophen. The 3-bromine can then be displaced by an amino group, and the 3-amino-2-bromothiophen can react with more [7] to give 3-aminothiophen. Evidence has been presented for this type of mechanism rather than an aryne mechanism.[2]

The great disadvantage of the use of strong bases to generate arynes is that the bases are usually good nucleophiles as well, so the products obtained are those of addition of the base to the aryne, unless there is a good nucleophile suitably placed within the molecule (Chapter 8). An alternative means of generating *o*-halogenophenyl anions, developed by Wittig and by Huisgen, is the reaction of *o*-dihalogenoaromatic compounds with lithium or with magnesium. Lithium alkyls can also generate arynes from the *o*-dihalogeno compounds, although of course these reagents are good nucleophiles and may react with the arynes.

Reactions of this sort have been used to generate many arynes, some of which, like 2,3-dehydropyridines[3] and tetraphenylbenzyne,[4] are not available by other routes.

A great advance in aryne chemistry was the discovery by Wittig that benzyne, generated this way, could be trapped by dienes in a Diels–Alder reaction. The reaction of arynes with dienes has since been regarded as a diagnostic test for the intermediates (but see ref. 5).

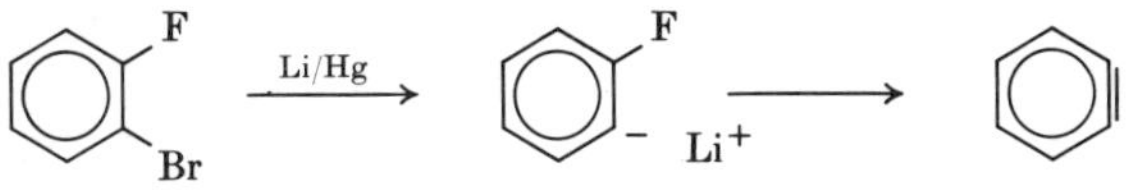

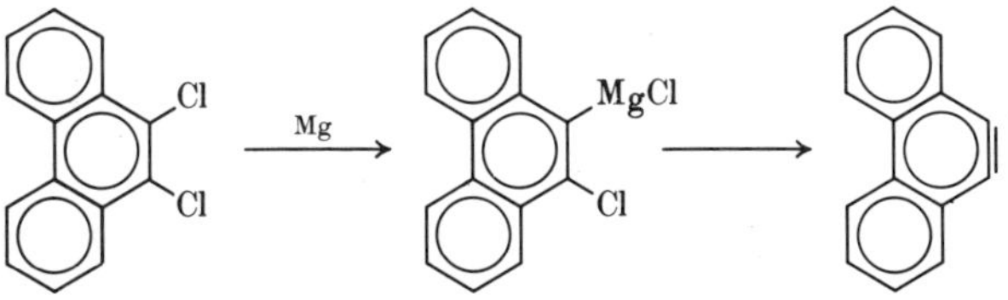

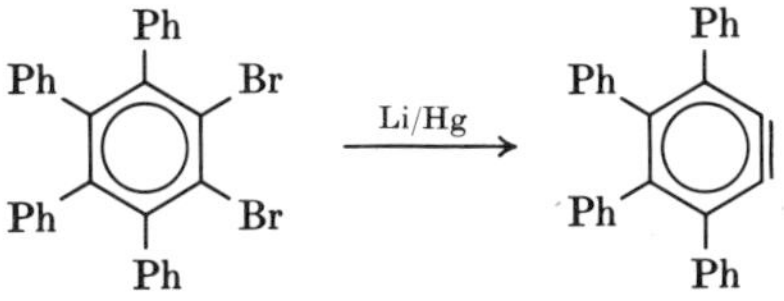

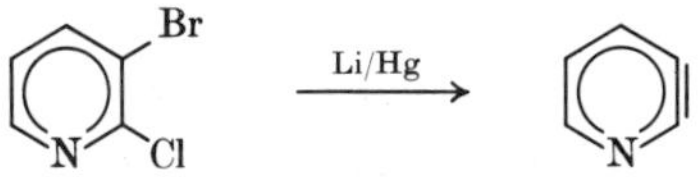

There is a close similarity between many of these routes to o-halogenophenyl anions and the routes to trihalogenomethyl anions used to generate dihalogenocarbenes. There are other parallel routes, too. For example, benzyne can be generated (in rather low yield) by the thermolysis of salts of o-halogenobenzoic acids, just as dichloro-carbene is formed from sodium trichloroacetate.

(M = Ag, Na, K, Cs;
X = F, Cl, Br, I)

Some other reactions that have been shown to give benzyne, probably via o-halogenophenyl anions, are outlined overleaf.

The last example is a commercially useful reaction that involves a benzyne intermediate: the Dow synthesis of phenol and diphenyl ether from chlorobenzene. If the carbon carrying the chlorine is labelled with ^{14}C, then the phenol formed in the reaction has ^{14}C at all the possible positions in the ring. This leads to an important

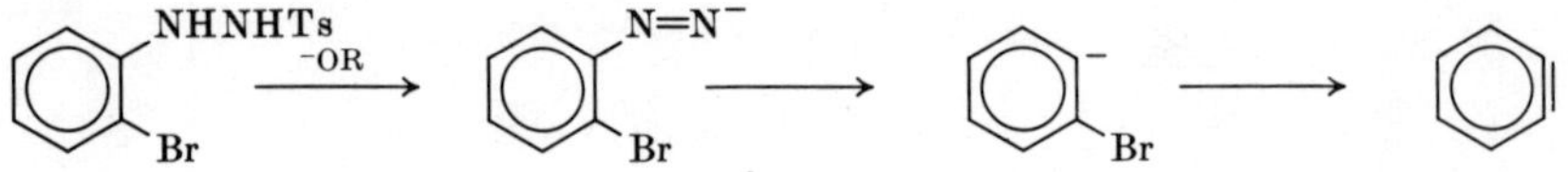

(ref. 6)

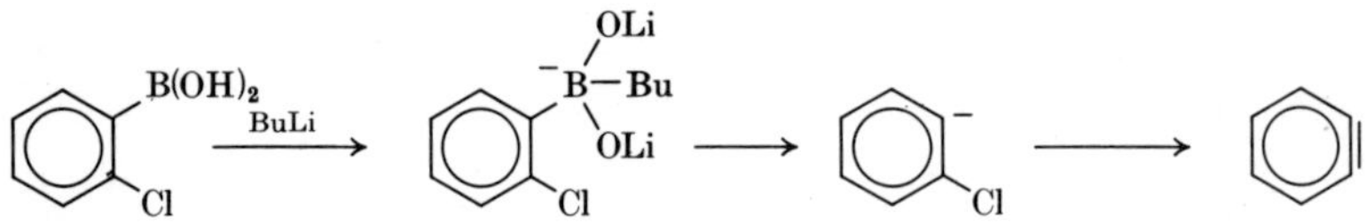

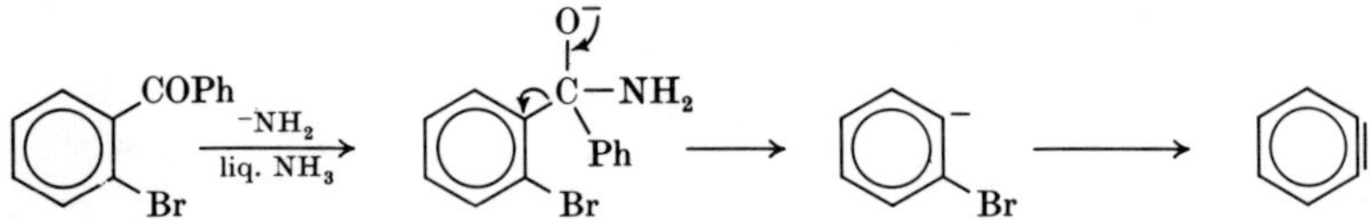

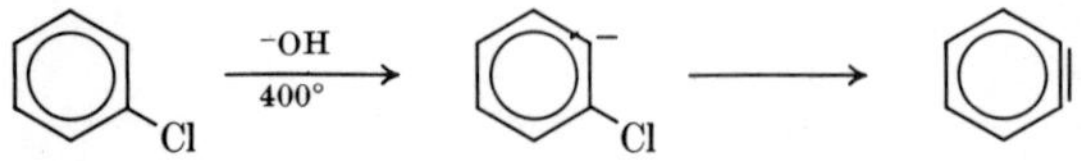

(ref. 7)

conclusion, which probably also applies to many of the other routes to benzyne that go through *o*-halogenophenyl anions: the loss of the halide from the anion is reversible:

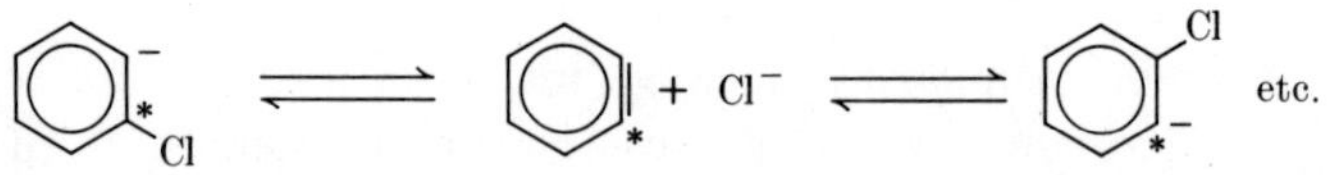

etc.

All the routes to arynes described so far involve halides as the leaving groups. Arynes have occasionally been made by routes involving other anions as leaving groups. The chlorate anion, ClO_3^-, is a good leaving group, and benzyne can be generated from perchlorylbenzene [8] and sodamide in liquid ammonia. However, other good leaving groups, such as benzoate and thiophenoxide, are also very good nucleophiles towards arynes, and so the anions from phenyl benzoate and diphenyl sulphide are not significantly cleaved to arynes.

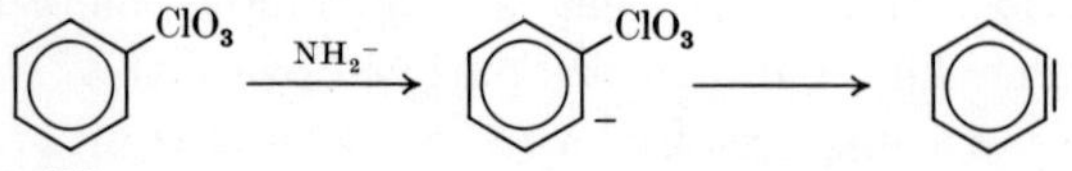

[8]

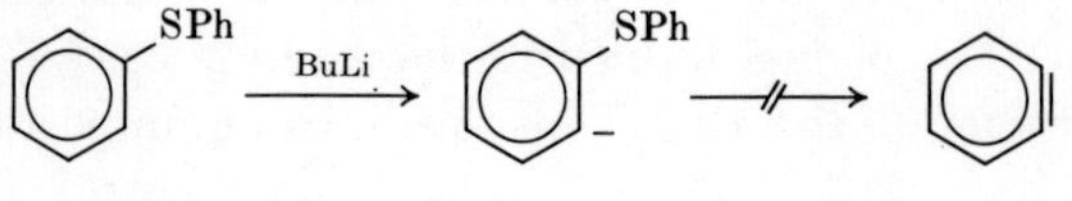

Other possible leaving groups are neutral molecules: nitrogen, tertiary amines, and phosphines, for example. These routes require zwitterionic intermediates, and are described in Section 4–4.

4–3 From aryl cations

It seems possible that arynes may be generated simply by diazo-tization of aromatic amines, if the cation formed by loss of nitrogen from the diazonium cation is sterically very crowded, and resistant to the normal nucleophilic attack. Then loss of the o-hydrogen as a proton becomes competitive. For example, diazotization of 2,5-di-t-butylaniline gives products that can can be rationalized by postulat-ing di-t-butylbenzyne as an intermediate:[8]

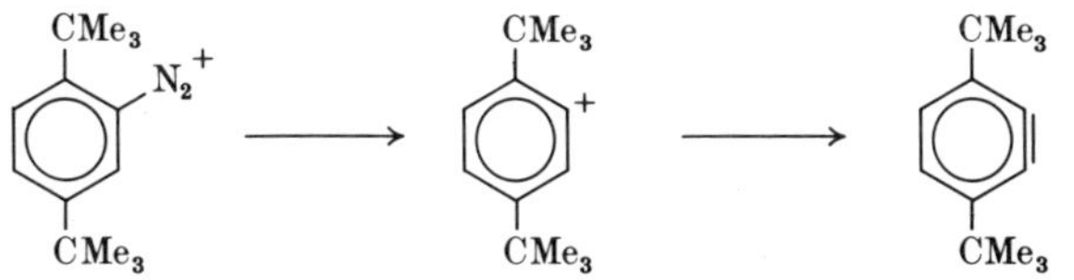

The thermal decomposition of derivatives of N-nitrosoacetanilide could also lead to the formation of benzynes via arylcarbonium ions, although the mechanism of these reactions is not clear:[9]

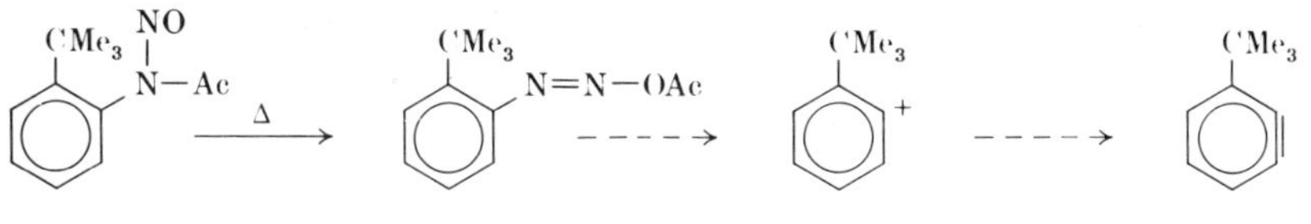

4–4 From zwitterions

Several important methods of generating arynes can be represented by the general scheme:

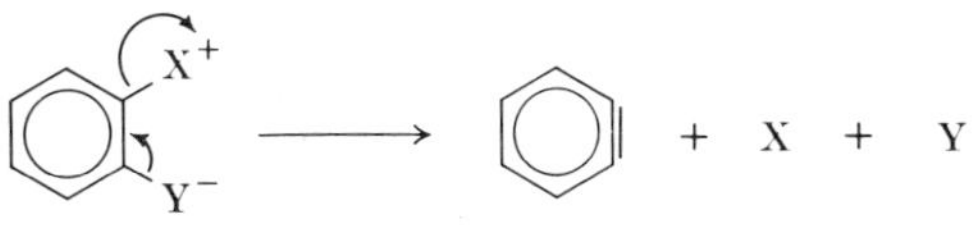

For such methods to be practicable, X and Y must both be stable molecules. The loss of X and Y need not be simultaneous; if it is not, aryl cations or anions may be intermediates.

The most important benzyne precursor of this sort is benzene-diazonium-2-carboxylate [9]. This compound is a very simple source of benzyne: it can be generated from anthranilic acid and decomposed *in situ* at temperatures below 80°, and can give benzyne adducts in high yield.[10] [Isolation of solid [9] is dangerous.] Substituted benzene-diazonium-carboxylates react similarly; the route has also been used to generate 3,4-dehydropyridine and 2,3-dehydronaphthalene. Flash photolysis of the zwitterion [9] has also been used to generate benzyne; by this means, the ultraviolet spectrum of benzyne was obtained.[11] Diphenyliodonium-2-carboxylate [10] is a similar zwitterionic benzyne precursor, but more stable. It decomposes in solvents at about 200° to benzyne, carbon dioxide, and iodobenzene. The decomposition of [9] is stepwise, with the carbon dioxide being lost after the nitrogen. Molecules such as the β-lactone [11] may be intermediates in the reaction.[12] The mode of decomposition of [10] is less clear.

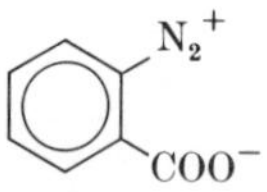

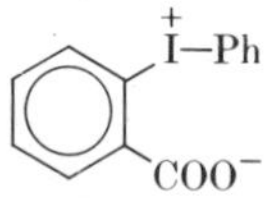

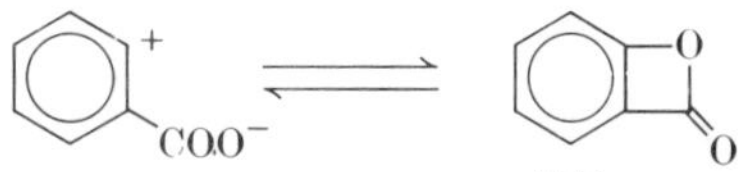

If the conditions of the thermolysis of [10] are not vigorous enough, there is an intramolecular rearrangement to phenyl 2-iodobenzoate. Similar reactions also occur when compounds such as benzenetri-methylammonium-2-carboxylate [12] and benzenedimethylsul-phonium-2-carboxylate [13] are heated, and no benzyne can be generated from them.[13]

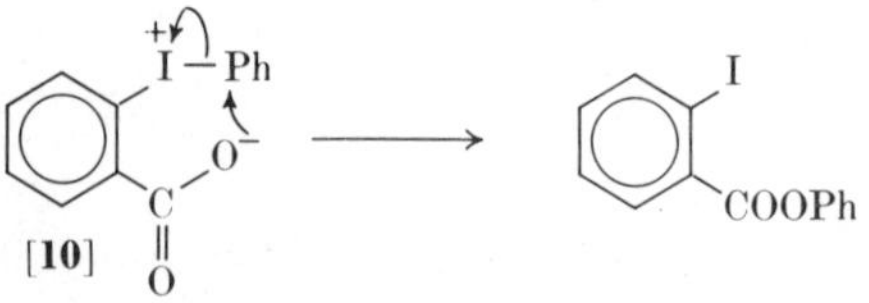

Other *ortho*-substituted benzenediazonium zwitterions have also been investigated as possible benzyne precursors, but without much success. Benzenediazonium-2-boronate [14] cleaves in mild conditions to give benzyne adducts in fair yields,[14] but benzenediazonium-2-sulphonate [15] gives no benzyne.

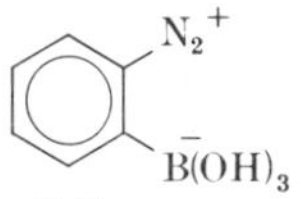

|14| 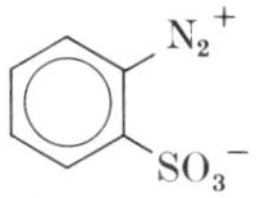 |15|

4–5 From aryl radicals

Benzyne is probably formed in the photolysis of o-di-iodobenzene, and o-iodophenyl radicals are likely intermediates. These radicals presumably give benzyne faster than they can abstract hydrogen intermolecularly, because in CH_3OD a product of the photolysis is o-deuteroanisole. An aryl radical would have abstracted an α-hydrogen atom from the solvent, to give anisole.[15]

The formation of benzyne from o-iodophenylmercuric iodide [16] and from phthaloyl peroxide [17] on pyrolysis may also involve aryl radicals as intermediates.

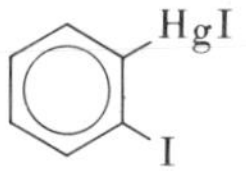

|16| 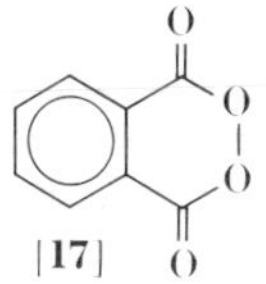 |17|

4–6 By fragmentation of cyclic systems

Any cyclic system that, in principle, could fragment by electronic rearrangement into benzyne and small stable molecules, like carbon monoxide and nitrogen, is a potential benzyne precursor. The conditions needed to generate benzyne from such a system may vary enormously, however; they will depend on the ground-state energy of the molecule and the energy of the transition state required for the fragmentation. A system that can fragment in extremely mild conditions is the nitrene [18]: this intermediate loses nitrogen very rapidly, even at -80°, when it is generated by oxidation of 1-aminobenzotriazole [19]. The same nitrene can also be generated by photolysis of the sodium toluene-p-sulphonyl derivative [20]. The

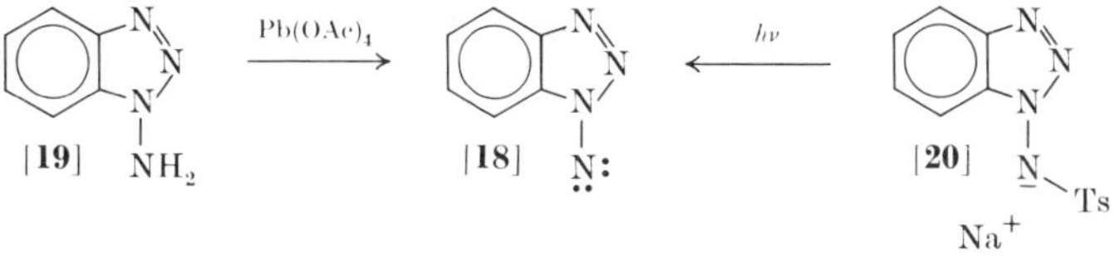

nitrene will clearly have high ground-state energy, and, since two stable nitrogen molecules are being formed in its fragmentation, the transition-state energy will be low; hence the mildness of the conditions. Benzyne is generated in high yield. The same route has been used to generate cycloalkynes[16] and the 'meta-aryne' 1,8-dehydronaphthalene [21].[17]

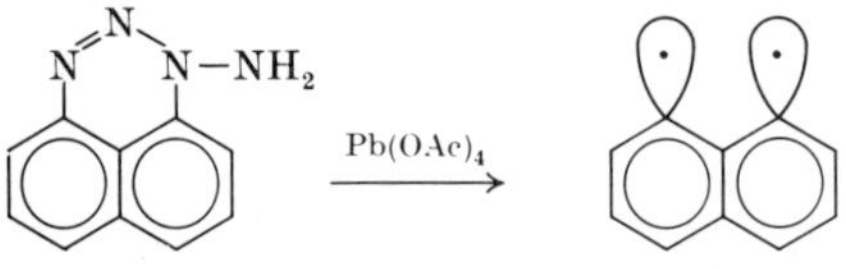

[21]

1,2,3-Benzothiadiazole 1,1-dioxide [22] is formed when aniline-2-sulphinic acid is diazotized. Although the compound can be isolated as a crystalline solid, it is unstable and fragments in very mild conditions to benzyne, nitrogen, and sulphur dioxide. This route has been extended to generate substituted benzynes, and 1,8-dehydronaphthalene.[18] With more stable molecules, the mass-spectral fragmentation pattern is often a good guide to the likely thermal breakdown pattern. This technique has been applied to relatively stable molecules like phthalic anhydride, indanetrione [23], and o-sulphobenzoic anhydride [24], all of which show a fragmentation to benzyne in the mass spectrometer and can be pyrolysed at temperatures in the region of 600–800° to give benzyne, usually in rather low yield.[19]

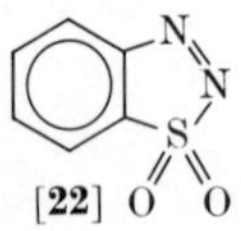

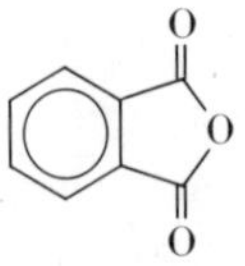

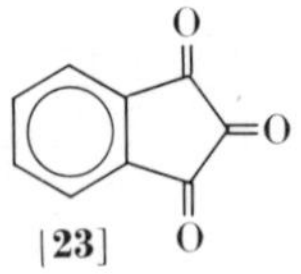

Benzyne intermediates generated by several different methods appear to be chemically indistinguishable. Benzyne, generated from three different precursors in the presence of two different dienes (furan and cyclohexadiene), gave the same ratio of adducts.[20]

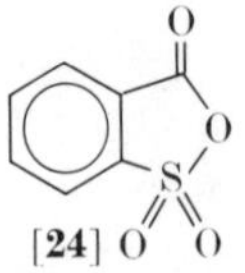

References

1. J. A. Zoltewicz and C. L. Smith, *J. Amer. Chem. Soc.*, 1966, **88**, 4766.
2. M. G. Reinecke and H. W. Adickes, *J. Amer. Chem. Soc.*, 1968, **90**, 511.
3. J. D. Cook and B. J. Wakefield, *Chem. Comm.*, 1968, 297; R. J. Martens and H. J. den Hertog, *Rec. Trav. Chim.*, 1964, **83**, 621.
4. D. Seyferth and H. H. A. Menzel, *J. Org. Chem.*, 1965, **30**, 649.

5. G. Wittig and E. R. Wilson, *Chem. Ber.*, 1965, **98**, 451.

6. J. F. Bunnett and D. A. R. Happer, *J. Org. Chem.*, 1966, **31**, 2369.

7. G. W. Dalman and F. W. Neumann, *J. Amer. Chem. Soc.*, 1968, **90**, 1601.

8. R. W. Franck and K. Yanagi, *J. Amer. Chem. Soc.*, 1968, **90**, 5814.

9. J. I. G. Cadogan and P. G. Hibbert, *Proc. Chem. Soc.*, 1964, 338; D. L. Brydon, J. I. G. Cadogan, D. M. Smith, and J. B. Thomson, *Chem. Comm.*, 1967, 727.

10. M. Stiles, R. G. Miller, and U. Burckhardt, *J. Amer. Chem. Soc.*, 1963, **85**, 1792; L. Friedman, *J. Amer. Chem. Soc.*, 1967, **89**, 3071.

11. R. S. Berry, G. N. Spokes, and M. Stiles, *J. Amer. Chem. Soc.*, 1962, **84**, 3570.

12. R. Gompper, G. Seybold, and B. Schmolke, *Angew. Chem. Internat. Edn.*, 1968, **7**, 389.

13. T. L. Gilchrist, C. W. Rees, and P. Serridge, unpublished observations.

14. L. Verbit, J. S. Levy, H. Rabitz, and W. Kwalwasser, *Tetrahedron Lett.*, 1966, 1053.

15. N. Kharasch and R. K. Sharma, *Chem. Comm.*, 1967, 492.

16. G. Wittig and H. L. Dorsch, *Annalen*, 1968, **711**, 46.

17. C. W. Rees and R. C. Storr, *Chem. Comm.*, 1965, 193.

18. R. W. Hoffmann and W. Sieber, *Annalen*, 1967, **703**, 96.

19. R. F. C. Brown and R. K. Solly, *Austral. J. Chem.*, 1966, **19**, 1045; E. K. Fields and S. Meyerson, *J. Org. Chem.*, 1966, **31**, 3307; *Advances Phys. Org. Chem.*, 1968, **6**, 1.

20. R. Huisgen and R. Knorr, *Tetrahedron Lett.*, 1963, 1017.

Problems

4–1 Suggest likely mechanism(s) for the following reaction:

(J. D. Roberts, D. A. Semenow, H. E. Simmons, and L. A. Carlsmith, *J. Amer. Chem. Soc.*, 1956, **78**, 601).

4–2 On heating with piperidine, 5-bromo-4-pyrimidone (A, R = H) gave 6-piperidino-4-pyrimidone (B) as the sole product, whereas the 6-methyl compound (A, R = Me) was inert to piperidine. Do these observations require the intermediacy of 5,6-dehydropyrimidone?

(R. Promel, A. Cardon, M. Daniel, G. Jacques, and A. Vandersmissen, *Tetrahedron Lett.*, 1968, 3067).

4–3   In the reactions

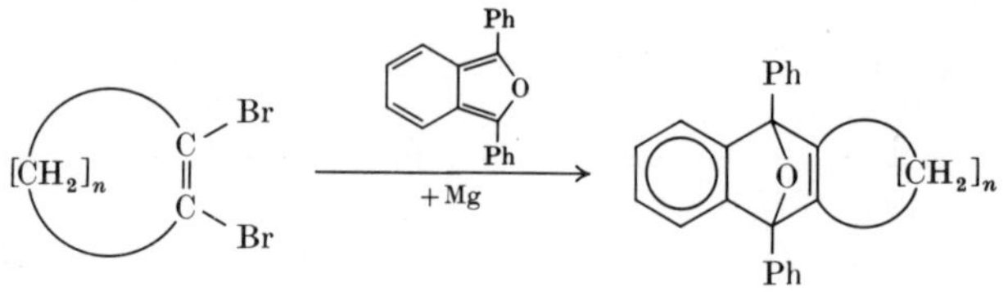

the yields of adducts are:

n:	6	5	4	3	2
%:	66	59	43	0·7	8

Comment on the probable mechanisms of formation of the adducts. How could the possible mechanisms be distinguished?
(Ref. 5).

4–4   When fluorenone and potassium hydroxide are heated in diphenyl ether, potassium biphenyl-2-carboxylate is formed. Similarly 3,6-dichloro-fluorenone gave 3′,5-dichlorobiphenyl-2-carboxylic acid (after acidification). However, 1,6- and 1,8-dichlorofluorenone gave, together with the expected biphenyl-2-carboxylic acids, the benzocoumarins A and B respectively. Suggest an explanation for this.

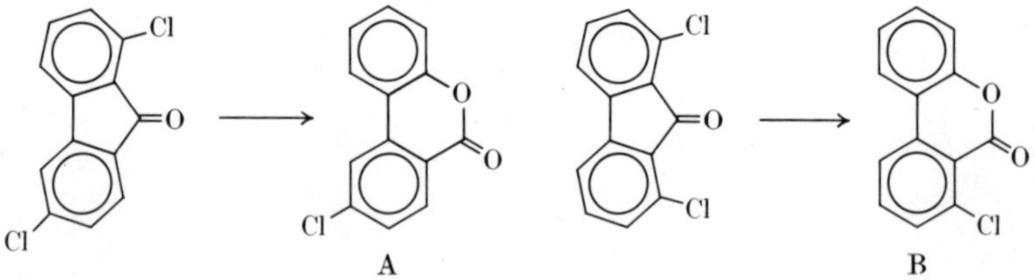

(See D. H. Hey, J. A. Leonard, and C. W. Rees, *J. Chem. Soc.*, 1963, 3125).

4–5   The following compounds have been used as benzyne precursors in our laboratory. Suggest the experimental conditions and likely mechanisms of formation of benzyne.

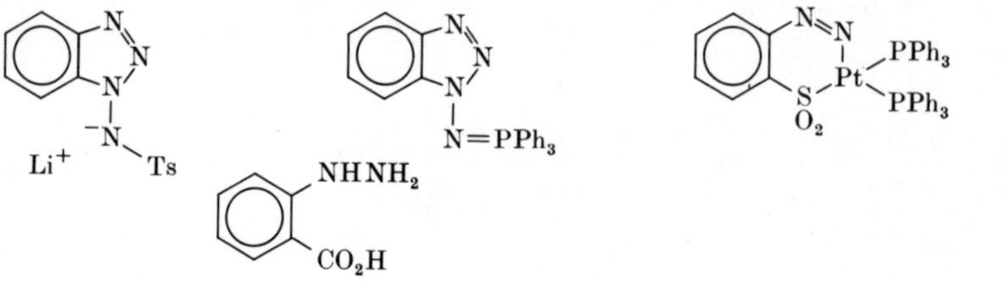

Cycloaddition reactions of carbenes

5–1 Introduction

One of the most characteristic reactions of carbenes is their addition to olefins to give cyclopropanes. The reaction, first reported by Doering and Hoffmann in 1954, has become an important method of synthesis of cyclopropanes and has been extended to other unsaturated systems, including allenes, acetylenes, and aromatic compounds.

In this chapter we shall consider the mechanisms and scope of the reaction. Some of its applications in organic synthesis are illustrated in Chapter 9.

5–2 Mechanism of the cycloaddition

A reasonable structure for the lowest singlet state of a carbene is an sp^2 hybrid with a vacant p orbital [1]. An alternative for the triplet state is a linear sp hybrid, with two singly occupied p orbitals [2].

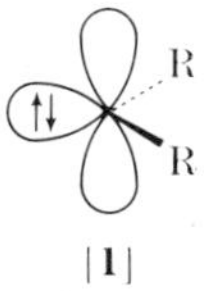

|1| 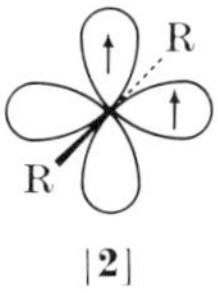 |2|

The similarity in structure between the singlet carbene [1] and a carbonium ion has been mentioned in Chapter 1, and it should extend to the reactions of the intermediates. Skell pointed this out in 1956 in some important work on the addition of dibromocarbene to olefins. Dibromocarbene behaved as an electrophile towards olefins, just as would be expected for an intermediate of structure [1]. In particular, the cycloaddition was stereospecific; that is, the stereochemistry of the olefin was retained in the cyclopropane formed in the reaction:

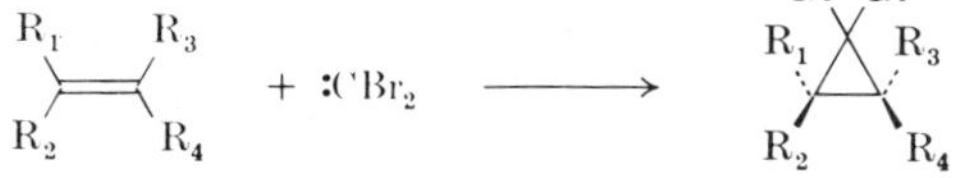

Skell rationalized this observation by pointing out that a singlet carbene could add to an olefin in a concerted manner, since the two new σ-bonds of the cyclopropane could be formed without changing the spin of any of the electrons involved. On the other hand, for a triplet carbene, the cycloaddition should go through a triplet diradical intermediate [3]. Before this diradical can close to a cyclopropane, there must be an inversion of spin of one of the electrons. If the free rotation about the C—C bonds of the diradical is faster than the spin inversion, then the stereochemistry of the original olefin will not be retained in the cyclopropanes:

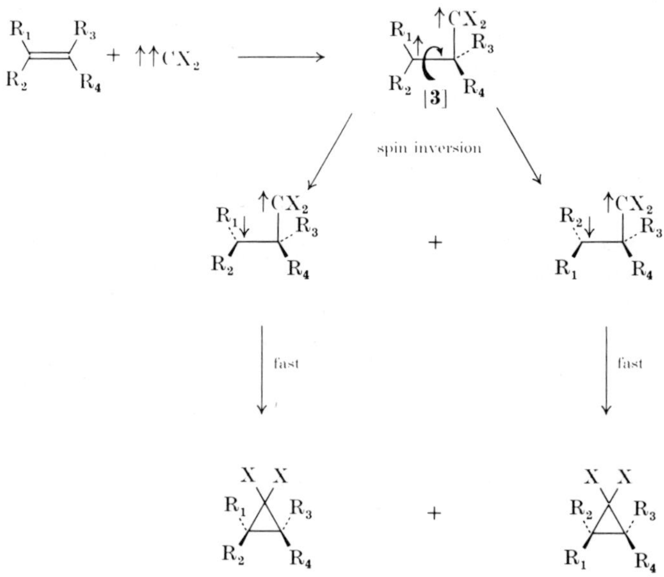

Indeed, there are many carbenes whose addition to olefins is not stereospecific, and it has often been possible to correlate stereospecific addition with carbenes in the singlet state, and non-stereospecific addition with triplet carbenes. The Skell theory has been valuable in rationalizing the stereospecificity of many carbene and nitrene reactions, even though it has been criticized on theoretical grounds.

A more sophisticated treatment of the cycloaddition of methylene to ethylene has been given by Hoffmann.[1] Woodward and Hoffmann have successfully applied the principle of conservation of orbital symmetry to cycloaddition reactions.[2] For this reaction, Hoffmann concludes that the bent singlet methylene approaches the ethylene initially through an unsymmetrical transition state [4]; the ethylene and methylene can then correlate with the ground state of cyclopropane and the cycloaddition is a 'symmetry-allowed' process. For methylene in the triplet state or the excited singlet state [5], a different geometry is predicted for the transition state, and the correlation is now, not with the ground state of cyclopropane, but with a triplet state of an excited configuration of the trimethylene diradical, $\cdot CH_2CH_2CH_2 \cdot$. In this, there are no barriers to rotation. The theory thus leads to the same results as the Skell theory for the addition of the lowest singlet and the triplet to olefins, although it is not primarily the spin state that determines the course of the addition, but the spatial distribution of the wave function. With this in mind, we can probably accept as a working hypothesis that carbenes in the *lowest* singlet state add stereospecifically to olefins, and carbenes in the triplet state do not.

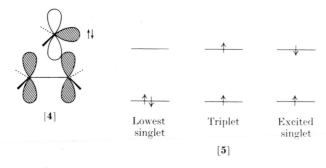

5–3 Cycloadditions to olefins

Since carbenes are electrophilic reagents, they react readily with electron-rich olefins, the relative rates of addition to different olefins being in the order of nucleophilicity of the olefins. Alkyl substituents increase the nucleophilic character of the olefins; cyclohexene, *cis*- and *trans*-butene, 2-methyl-2-butene, and 2,3-dimethyl-2-butene are commonly used as traps for carbenes. Olefins with a hetero atom adjacent to the double bond, such as enol ethers, enamines, and vinylic esters, are also very good carbene traps, although with these the mechanism of addition may involve prior co-ordination to the carbene through a lone pair of electrons on the hetero atom. The selectivity of the carbenes for different olefins depends on their internal structure and on the amount of excess energy they have. The influence of structure is shown by the relative selectivities of the dihalogenocarbenes, which are in the order of their stabilities, namely $:CF_2 > :CCl_2 > :CBr_2$. The method of generating carbenes determines the amount of excess energy they have: those generated by photolysis of diazoalkanes or by similar routes are generally much less discriminating in their reactions than those produced by base-induced α-eliminations, for example.

Steric factors may also influence the selectivity of carbenes for different olefins; for example, the effect of successive α-methylation

on the addition of dichlorocarbene to 1-butene is shown by the following relative rates:[3] $EtCH\!=\!CH_2$, 1·00; $Pr^iCH\!=\!CH_2$, 0·43;
$Bu^tCH\!=\!CH_2$, 0·029.

If carbenes produced by highly energetic routes add to olefins to
give non-stereospecific products, these products may have been
formed either by non-stereospecific addition of the carbenes, or by
isomerization of 'hot' molecules of the primary adduct. Before any
conclusions can be drawn about the mechanism of a carbene addition
it is necessary to rule out this latter possibility with a blank experiment. If the cyclopropanes can isomerize in the conditions needed to
generate the carbenes, then the final product composition will not
represent the initial carbene adducts. If the products do not isomerize, then Skell's theory can be tested. The correlation between
spin state and stereospecificity is usually very good. Carbenes like
dihalogenocarbenes, for which the lowest singlet is especially stable,
always add stereospecifically, however they are generated. With
methylene, phenylcarbene, fluorenylidene, and others that have a
triplet ground state, the results of addition to olefins are very instruc-
tive. Photolyses of diazoalkanes and similar methods of generation
are likely to produce carbenes in a singlet state, so, in the presence of
olefins, cyclopropanes can be formed as follows:

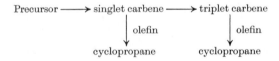

There are various ways in which a system of this sort can be
investigated. One is to vary the concentration of the olefin in the
reaction mixture: the lower the concentration, the more likely it is
that the carbenes will be in the ground state when they react,
because of the increased opportunity for deactivating collisions.
With carbenes known to have a triplet ground state, the effect of
lowering the olefin concentration is always to make the addition *less*
stereospecific, as the Skell theory would predict. Instead of altering
the olefin concentration, the solvent can be replaced with one that
is much better at deactivating the excited carbenes. Dibromomethane
and hexafluorobenzene are good solvents for this purpose. This has
the same effect as lowering the olefin concentration: more of the
carbenes are in the triplet ground state before they react with the
olefin, and the addition is therefore less stereospecific. A third
type of experiment is to add a trap that will react selectively
with the triplets. The effects of adding such traps is to make the

addition of the olefins *more* stereospecific, since it is mainly the singlet carbenes that add. Suitable traps include olefins which can form very stable radicals, because the addition of triplet carbenes to olefins involves diradical intermediates. Styrene, α-methylstyrene, and butadiene have been used, because they react very much faster with triplet carbenes than with singlets.

A reaction where techniques of this sort have been used is the photolysis of diazofluorene in the presence of *cis*-butene to give the *cis*- and *trans*-cyclopropanes.[4] In the butene as solvent, the product is mainly the *cis*-cyclopropane with some of the *trans*-isomer; this suggests that both singlet and triplet carbenes are adding to the *cis*-butene. If the reaction is carried out in hexafluorobenzene containing *cis*-butene, the proportion of the *trans*-cyclopropane increases, because more of the addition is due to triplet carbenes. On the other hand, if the photolysis is carried out in *cis*-butene containing butadiene, the proportion of *cis*-cyclopropane increases, because the butadiene selectively removes the triplet carbenes.

Me
Me
Me
Me
Me
Me

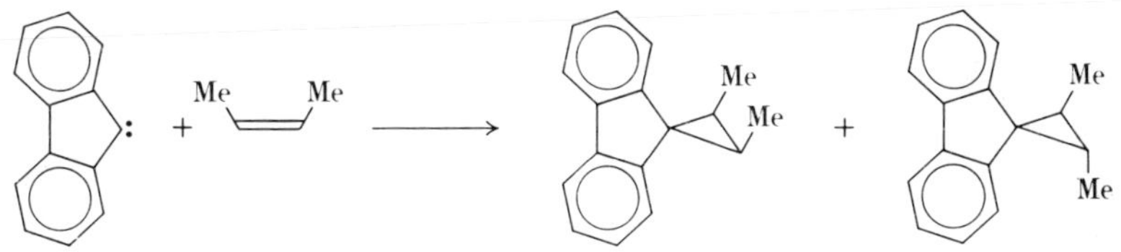

in *cis*-butene: 2 1
in hexafluorobenzene and *cis*-butene: proportion of *trans* increases
in *cis*-butene containing butadiene: proportion of *cis* increases

If triplet carbenes add to conjugated dienes to give resonance-stabilized diradical intermediates, some 1,4-cycloaddition might be expected as well as the normal 1,2-addition:

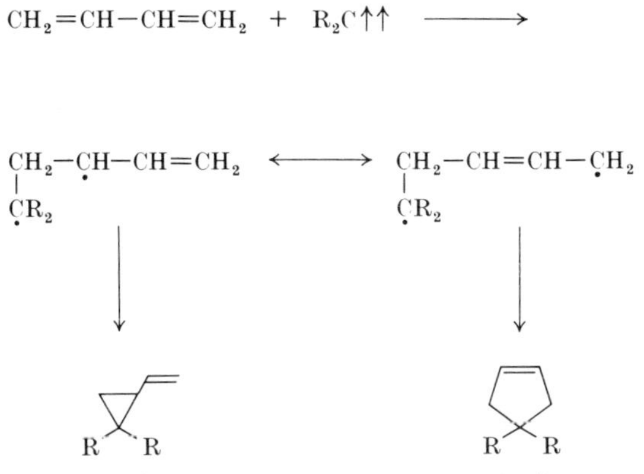

CH₂=CH—CH=CH₂ + R₂C↑↑
CH₂—CH—CH=CH₂
CR₂
CH₂—CH=CH—CH₂
CR₂
R R
R R

(1,2) (1,4)

In fact, there are hardly any examples of 1,4-addition by carbenes. One is the addition of phenylcarbene, presumably the triplet, across the 9- and the 10-position of anthracene, in about 1% yield:[5]

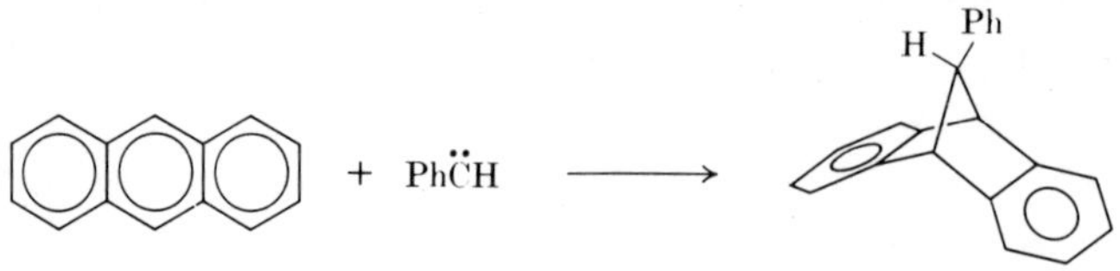

Thus the general picture of carbene addition to olefins is a ready addition to nucleophilic olefins by both singlet and triplet carbenes, and a very fast addition by triplet carbenes to conjugated dienes and other olefins that can give stable radical intermediates. The stereochemistry of the addition fits in very well with the Skell theory. The chief exceptions to this pattern are carbenes that rearrange too rapidly to enable them to be trapped (simple alkylcarbenes, for example), and a few other carbenes that we can describe as nucleophilic. These are carbenes in which the singlet structure is stabilized by incorporating the vacant p orbital into some delocalized electronic system, so that the electrophilic character of the carbenes is suppressed. Intermediates with a nitrogen or oxygen adjacent to the bivalent carbon are in this category. Such carbenes may not add to cyclohexene and to similar olefins. Two other interesting examples are diphenylcyclopropenylidene [6] and cycloheptatrienylidene [7]. The singlet structures of these can be stabilized through aromatic resonance forms in which the vacant p orbital of the carbene carbon becomes part of an aromatic system.[6] These carbenes react with dimethyl

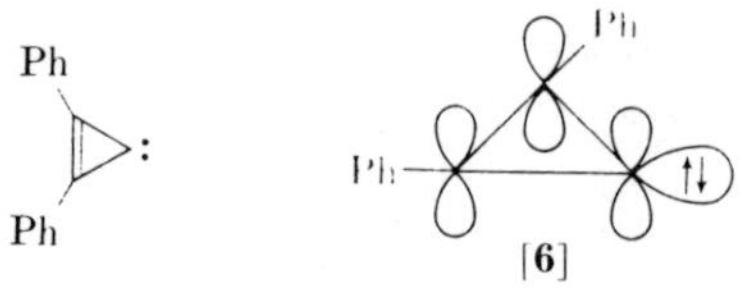

[6]

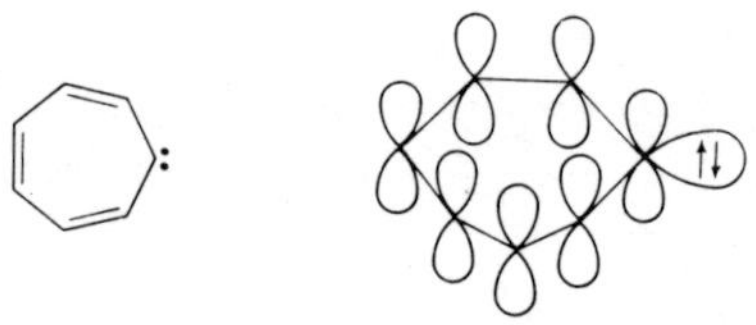

[7]

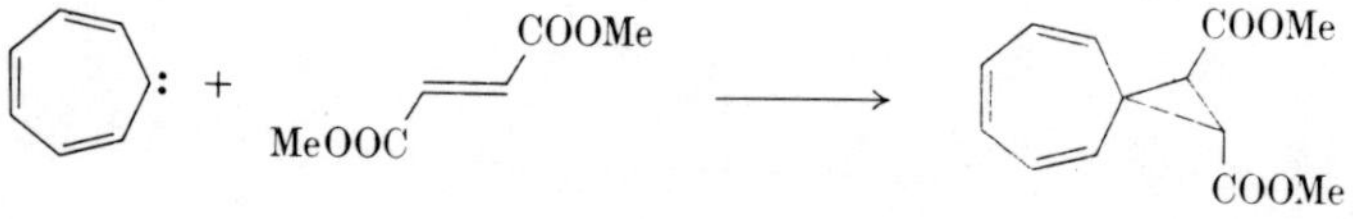

fumarate, an electrophilic olefin, to form the cyclopropanes stereospecifically. Olefins like dimethyl fumarate do not form cyclopropanes with normal carbenes.*

5–4 Cycloadditions to allenes and acetylenes

The reaction between carbenes and allenes, to give alkylidenecyclopropanes, is a straightforward one. Dihalogenocarbenes add readily to alkyl-substituted allenes at the more substituted double bond.[7] Methylene, from diazomethane, adds to allene to give methylenecyclopropane, and to methylallene to give the two possible alkylidenecyclopropanes:

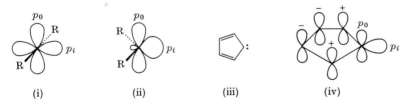

* Again, a more sophisticated treatment of this problem has been given by Hoffmann, and this is briefly described here.

The non-bonding orbitals, p_0 and p_i, in a linear carbene (i) are degenerate; that is, they are of equal energy. A singlet ground state is only possible for the carbene if the energy levels of these two orbitals are separated by about 2 eV (about 50 kcal. mole^{-1}). One way in which the p-orbitals can become non-degenerate is by making the carbene non-linear, as in (ii); the p_i orbital in the bending plane is stabilized and acquires s character, while the p_0 orbital perpendicular to the bending plane is unchanged in energy. Another way is by selective interaction of the p_0 orbital with a π-electron system containing either a high lying occupied level, or a low lying unoccupied level of the correct symmetry. Interaction of p_0 with a high lying occupied level will tend to destabilize p_0, but interaction with a low lying vacant orbital will stabilize it.

In a conjugated $4n\pi$ electron system, the orbital with the correct symmetry for interaction with p_0 is the lowest unoccupied one; for a $(4n + 2)\pi$ system it is the highest occupied one. Thus:

out-of-plane orbital p_0
   (a) unaffected by bending.
   (b) destabilized by interaction with a high lying filled π level, as with $(4n + 2)\pi$ systems.
   (c) stabilized by interaction with a low lying unfilled π level, as with $4n\pi$ systems.

in-plane orbital p_i
   (a) stabilized by bending.
   (b) unaffected by a π system orthogonal to it.

(i)   (ii)   (iii)   (iv)

For example, in *cyclopentadienylidene* (iii) p_0 is stabilized by interaction with the lowest unoccupied π orbital of the diene, as in (iv). p_i is also stabilized by bending, however, and there is hardly any difference in the resulting energy levels of the two orbitals. The carbene (iii) therefore has a triplet ground state.

In *cyclopropenylidene* [6] and *cycloheptatrienylidene* [7] p_0 is destabilized by interaction with the highest filled orbitals of the 2π or 6π system, respectively. p_i is stabilized by bending so that the energy level of p_i is below that of p_0. In cyclopropenylidene the energy difference is more than 3 eV, so this carbene should have a singlet

Ethoxycarbonylcarbene has also been added to allene, to give the ethyl ester of 2-methylenecyclopropane-1-carboxylic acid. This type of cycloaddition is of interest mainly because it provides a potential synthetic route to alkylidenecyclopropanes that can be very difficult to make by other means.

The addition of carbenes to acetylenes is similarly important, because it can be a good way of synthesizing cyclopropenes. So far it has not had much application in synthesis, because the yields of cyclopropenes tend to be low; like other electrophilic additions, carbene addition to acetylenes is, in general, more difficult than to olefins. Another complication is that acyl- and alkoxycarbonyl-carbenes can add to acetylenes in a 1,3-dipolar manner, to give furan derivatives:

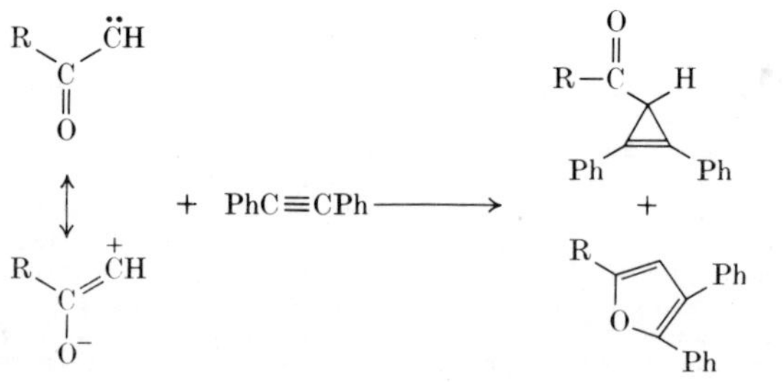

The furans may also be formed by isomerization of the cyclo-propenes, rather than by direct 1,3-dipolar addition. Benzoylcarbene adds to diphenylacetylene to give mainly 2,3,5-triphenylfuran; ethoxycarbonylcarbene gives mixtures of the cyclopropene and the furan, the proportions depending on the reaction conditions. Dihalogenocarbenes add normally to give the dihalogenocyclopro-penes, though these may be formed only in low yield. As with olefins, the reaction goes better with electron-releasing substituents on the acetylenes. The dihalogenocyclopropenes are intermediates in the synthesis of cyclopropenones, several of which have been made this way:[8]

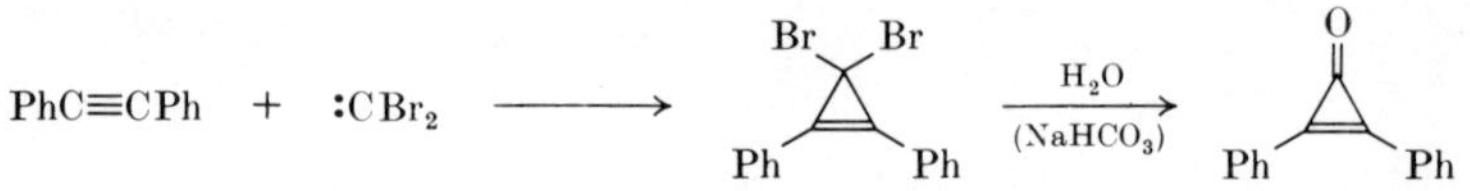

---

ground state. Though the energy difference is not so great for cycloheptatrienylidene, this too may have a singlet ground state. These carbenes should be nucleophilic because of the increased electron density transferred to p_0 by interaction with the filled π levels.

(R. Gleiter and R. Hoffmann, *J. Amer. Chem. Soc.*, 1968, **90**, 5457).

If a molecule contains both an olefinic bond and an acetylenic bond, then we might have expected the olefinic bond to be attacked preferentially. This is not always so: an investigation of dichlorocarbene addition to conjugated enynes showed that the triple bond was attacked preferentially in several systems:[8]

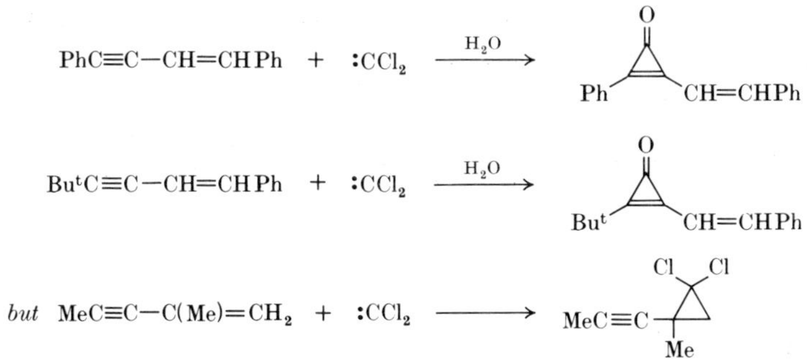
PhC≡C—CH=CHPh + :CCl2
H2O
O
Ph
CH=CHPh
ButC≡C—CH=CHPh + :CCl2
H2O
O
But
CH=CHPh
but MeC≡C—C(Me)=CH2 + :CCl2
Cl Cl
MeC≡C
Me

5–5 Cycloadditions to aromatic systems

The addition of carbenes to aromatic systems, to give the products of ring-expansion, is an important reaction in organic synthesis. In 1885, Buchner and Curtius discovered that ethyl diazoacetate and toluene reacted on heating to give an ester that was an isomer of ethyl 3-phenylpropionate [**8**].

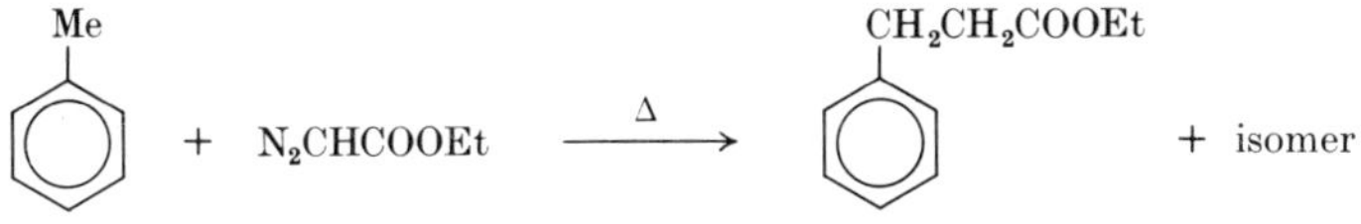
Me
+ N2CHCOOEt
Δ
CH2CH2COOEt
+ isomer

[**8**]

By the end of the century Buchner had very thoroughly investigated the reaction between benzene or toluene and ethyl diazoacetate, and had discovered the nature of the products; they were cycloheptatriene esters. The cycloheptatriene acids are still sometimes called 'Buchner acids' in recognition of his work. We would now rationalize Buchner's reaction as an attack of ethoxycarbonylcarbene on the aromatic system to give a norcaradiene ester [**9**] which can ring-expand by a Cope rearrangement to the cycloheptatriene esters [**10**].

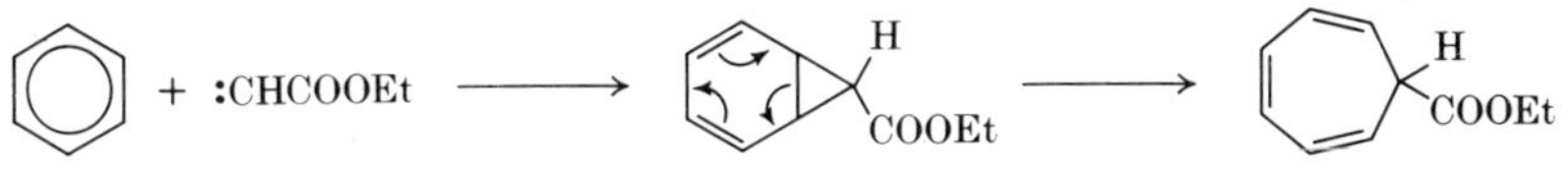
+ :CHCOOEt
H
COOEt
H
COOEt

[**9**] [**10**]
and isomers

This type of reaction will only take place with benzene derivatives if the carbenes are formed by a highly energetic route, namely thermolysis or photolysis of diazoalkanes, or similar methods. The norcaradienes cannot usually be isolated and the. isomeric cycloheptatrienes are produced, along with products of insertion into alkyl side chains on the aromatic system, if there are any. The latter reaction can be suppressed by using copper catalysts. Examples of the reactions with benzene are shown:[9,10]

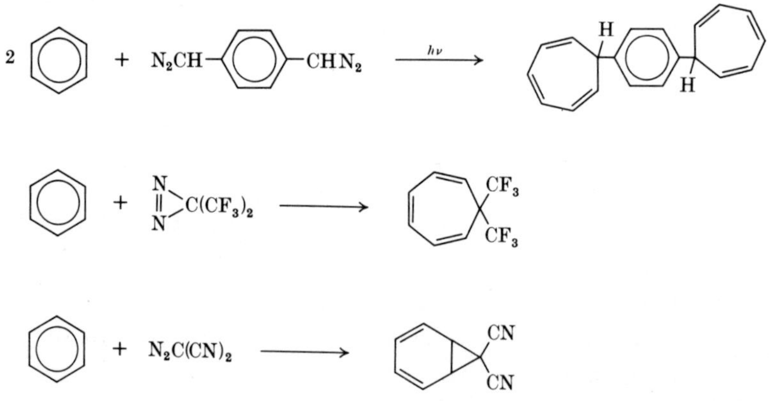

It is interesting that, with dicyanocarbene, the norcaradiene structure is the stable one, rather than the cycloheptatriene.

Most other aromatic systems can give similar cycloaddition reactions with carbenes. The reaction is progressively easier as the bonds that are attacked have more and more double-bond character. It is also facilitated by electron-releasing substituents, especially alkoxy groups, on the double bond. Thus, the 1,2-bond of naphthalene is much more reactive towards carbene addition than the 2,3-bond. The 9,10-bond of phenanthrene and the bonds of the central ring of anthracene are attacked more readily still. In some of the adducts, the norcaradiene structure is the stable one, and in others it is the cycloheptatriene structure. The differences can usually be simply rationalized on the basis of gain or loss of resonance energy. For example, the adduct across the 1,2-bond of naphthalene exists in the norcaradiene structure [11] because the unsubstituted ring can be fully aromatic, whereas the 2,3-adduct exists entirely as the cycloheptatriene [12] for the same reason.

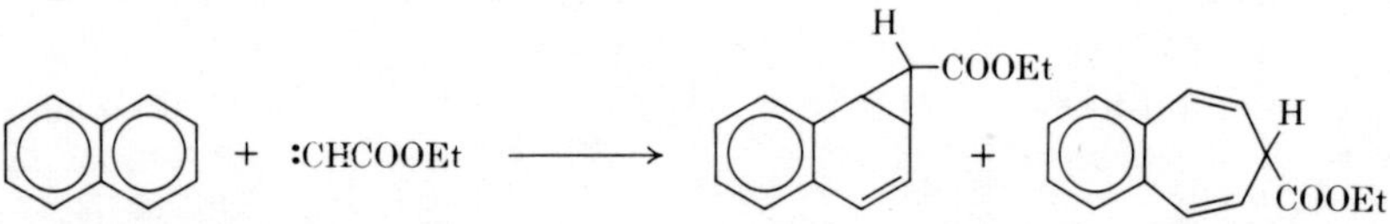

As the double-bond character increases it becomes possible to add less energetic carbenes, such as the dihalogenocarbenes. Thus, dichlorocarbene will add to the 1,2-bond of naphthalene if the bond is activated by a methoxyl substituent. 9-Methoxyphenanthrene reacts similarly, and in this case the dichlorocyclopropane can be isolated:

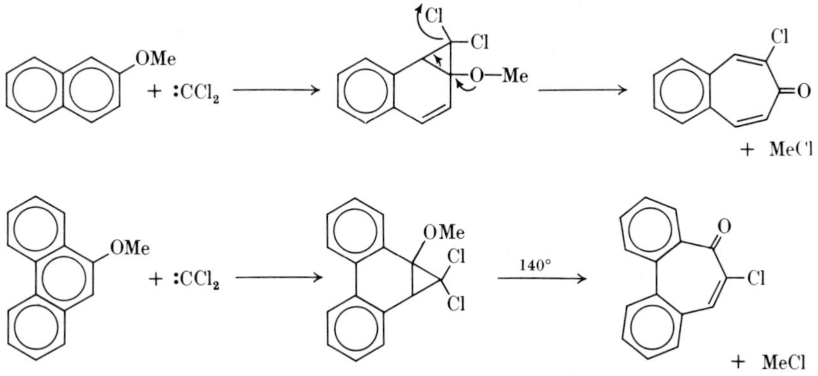

Dichlorocarbene generated from chloroform and ethylene oxide will add to unsubstituted anthracene, phenanthrene, and naphthalene.[11]

Monochloro- and monobromo-carbenes (or the carbenoids) are reactive enough to attack benzene. The products are stable tropylium salts:

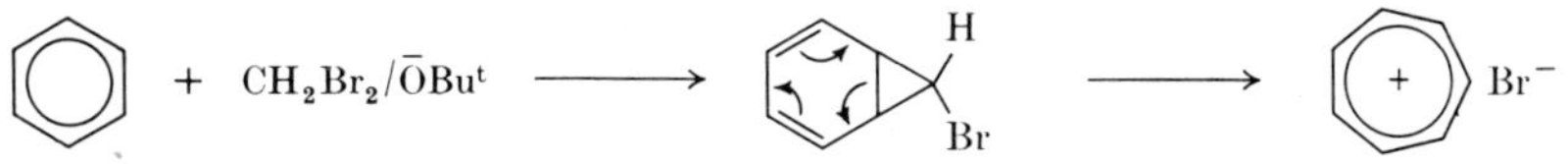

Reactions of this sort can also be carried out with heterocyclic aromatic systems. Pyrroles, indoles, pyrazoles, and similar systems add carbenes readily to give ring-expanded products. These reactions are sometimes useful synthetically, especially with dihalogenocarbenes.

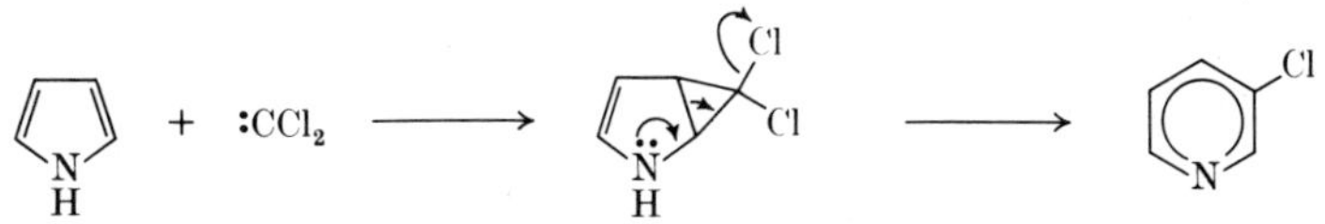

With pyrroles and related heterocyclic systems, there are two possible modes of attack by dihalogenocarbenes in basic conditions: an attack on a double bond of the neutral molecule, as above, or an attack at some nucleophilic site of the negatively charged conjugate base. The latter reaction leads to substitution products, and is

exemplified by the Reimer–Tiemann reaction of pyrrole with chloroform and a base. A mechanism for the Reimer–Tiemann reaction is
outlined in the next Chapter. Both ring-expanded and substitution
products can be found together in reactions of these heterocyclic
compounds with dihalogenocarbenes; 2,3-dimethylindole reacts with
dichlorocarbene as shown:[12]

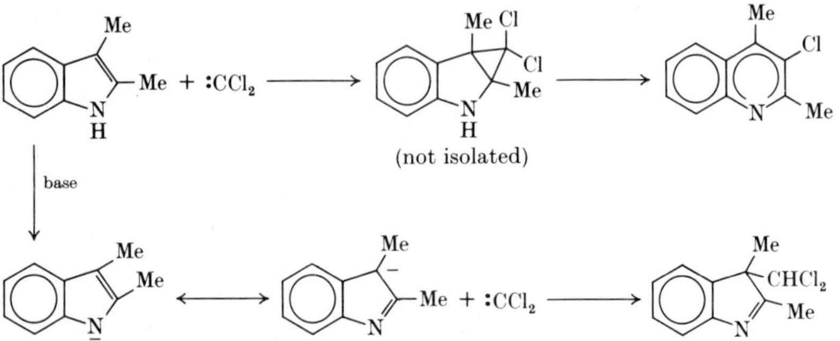

(not isolated)

5–6 Cycloadditions to other unsaturated systems

In principle, the cycloaddition reactions of carbenes need not be
limited to $\diagup{C}{=}{C}\diagdown$ double bonds; similar reactions might be possible
with $\diagup{C}{=}O$, $\diagup{C}{=}S$, $\diagup{C}{=}N{-}$, $-N{=}N{-}$, $-C{\equiv}N$, and others. If
cycloadditions were possible in such systems, they would probably
go by a different mechanism, since the hetero atom has a lone pair
of electrons available for prior co-ordination to the carbene:

$$\diagup{C}{=}X: + \;:CR_2 \longrightarrow \diagup{C}{=}\overset{+}{X}{-}\overset{-}{C}R_2 \longleftrightarrow \diagup\overset{+}{C}{-}X{-}\overset{-}{C}R_2$$

Other mechanisms are available for forming cycloadducts with
polarized double bonds. For example, diazoalkanes can add as 1,3-
dipoles to give unstable heterocyclic compounds that collapse to
three-membered-ring products:

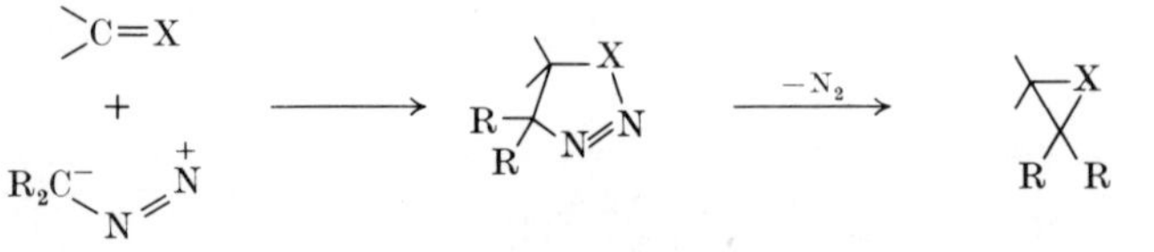

Oxocarbenes can themselves act as 1,3-dipoles and add in a similar
fashion:

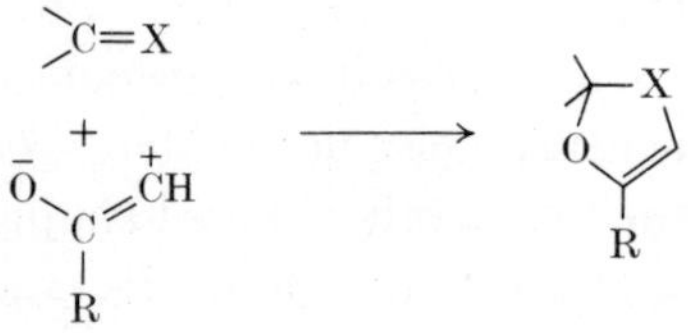

There are examples of cycloadditions to $\diagdown$C=O, $\diagdown$C=S, and

$\diagdown$C=N— bonds that probably involve carbene intermediates, and

cycloadditions to $\diagdown$C=O and to —C≡N that involve oxocarbenes
in the dipolar form. As yet none of these reactions appear to be very
general, however. Some examples of reactions of this sort are shown:

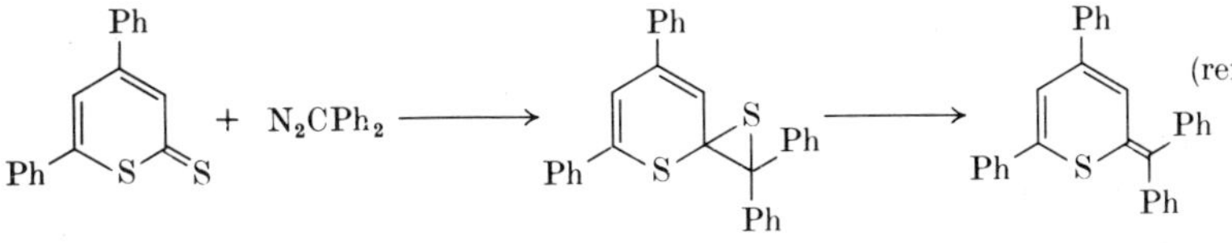

(ref. 13)

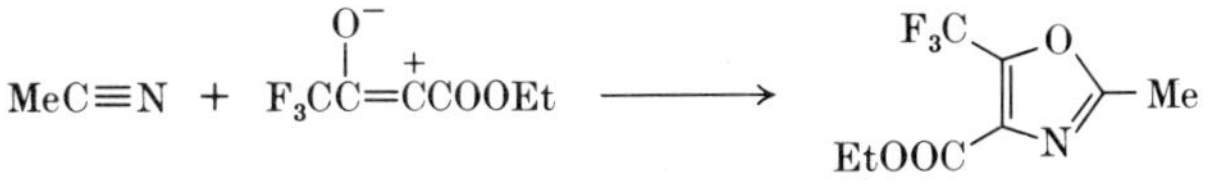

$MeC≡N + F_3C\overset{O^-}{\underset{+}{C}}=CCOOEt \longrightarrow$ (ref. 14)

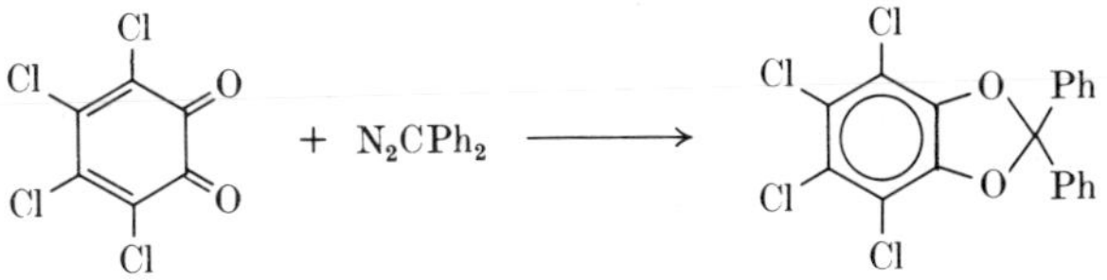

(ref. 15)

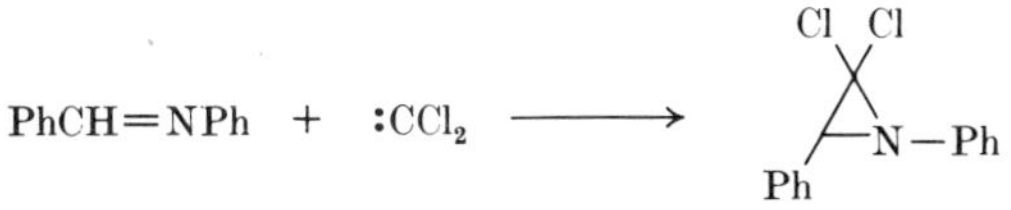

$PhCH=NPh + :CCl_2 \longrightarrow$ (ref. 16)

5–7   Carbenoid cycloadditions

Catalysed decomposition of diazo compounds. Copper and other
transition-metal catalysts can profoundly alter the product composition from diazoalkane reactions. The intermediates are more selective
than the free carbenes; cycloadditions to olefins are much cleaner and
go in higher yields. The addition to olefins is stereospecific. With an
unsymmetrical carbene like ethoxycarbonylcarbene, there may be
two possible stereoisomers formed with cyclohexene and similar
olefins:

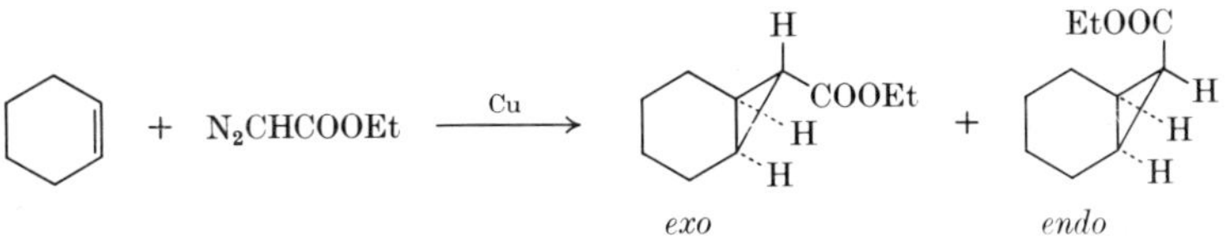

In copper-catalysed reactions, it is the sterically less crowded *exo*
isomer that is formed preferentially, indicating that the intermediate
is probably rather bulky, as a metal complex would be. Copper

catalysts can also enable acylcarbenes to add to olefins; in the absence of a catalyst the major reaction of a diazoketone is the Wolff rearrangement to a keten. Catalysed diazoalkane decompositions are thus very useful in cyclopropane synthesis.

Simmons–Smith reaction. The reagent formed between zinc and diiodomethane in ether or in 1,2-dimethoxyethane adds readily to nucleophilic double bonds to give cyclopropanes.[17] The addition is stereospecific. A comparison of the rates of cyclopropane formation with different olefins indicates that the intermediate is electrophilic, but sterically rather bulky. For example, the observed rates of addition of methylene, from the Simmons–Smith reagent, to cyclohexene and to tetramethylethylene, are almost equal. Presumably the greater nucleophilic power of the latter olefin is counterbalanced by its greater steric crowding. By comparison, dichlorocarbene reacts with tetramethylethylene about 50 times as fast as with cyclohexene, because steric factors are not so important with the free carbene.

The mechanism of cyclopropane formation is not certain, though it is probably a concerted cycloaddition. The infrared stretching frequency of the olefinic bond is not altered by the addition of methylzinc iodide, and this seems to rule out any extensive prior coordination of the olefin to the metal. A transition state such as the one shown is possible:

The ether solvent is certainly important in stabilizing the reagent. If the olefin has an oxygen function suitably placed near to the double bond, then the addition of methylene is very fast and leads to only one isomer even if several are possible. This may be because the oxygen can co-ordinate to the zinc in much the same way as that of the ether solvent:

The advantages of the Simmons–Smith reagent in synthesis are its convenience and its great selectivity compared with other sources of methylene.

α-Halogenolithium alkyls.[18] These can form cyclopropanes with olefins. The reagents are electrophilic, and the cyclopropanes are formed stereospecifically. Unlike many other carbenoid reactions, there seem to be no marked steric requirements for addition. The reaction may not be a good synthetic route to cyclopropanes, however, because there are so many side reactions of the organolithium intermediate that can compete with its addition to olefins. Among these are rearrangement, reaction with excess of the lithium alkyl used to prepare the α-halogenolithium intermediate, and dimerizing α-elimination to give olefins. Examples are shown of the reactions of trichloromethyl-lithium and of α-bromonorcaryl-lithium with cyclohexene:

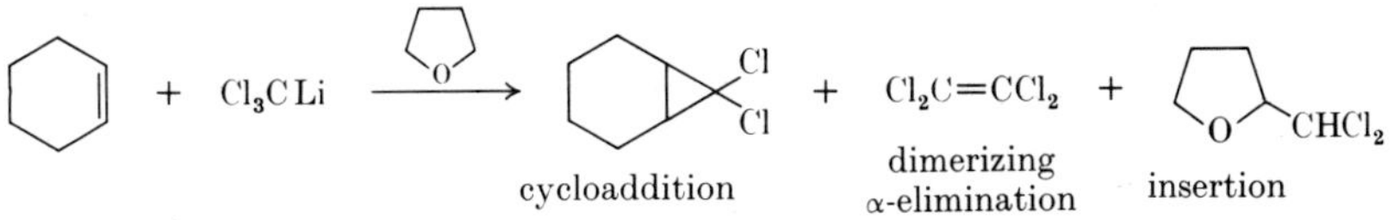

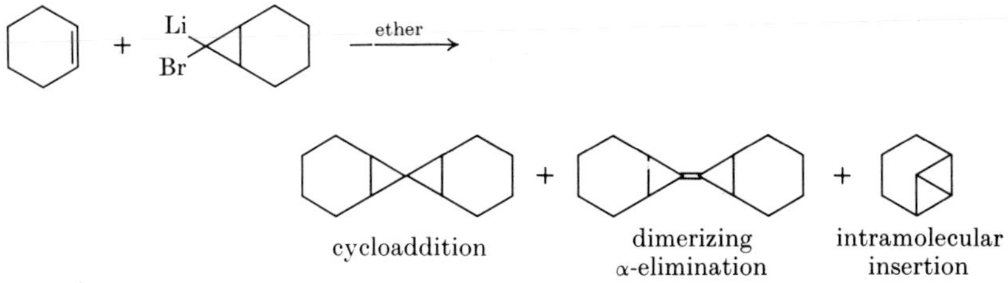

With each cycloaddition there are competing reactions.

Closs has compared the carbenoid addition to olefins of α-halogeno-lithium alkyls with addition of the free carbenes from the corresponding diazoalkanes.[19] Apart from the increased selectivity of the carbenoids, the most obvious difference is the stereospecificity of their reactions, even when the addition of the free carbene is completely non-stereospecific, as with phenylcarbene, for example.

References

1. R. Hoffmann, *J. Amer. Chem. Soc.*, 1968, **90**, 1475; see also A. G. Anastassiou, *Chem. Comm.*, 1968, 991.
2. R. Hoffmann and R. B. Woodward, *Accounts of Chemical Research*, 1968, **1**, 17.
3. R. A. Moss and A. Mamantov, *Tetrahedron Lett.*, 1968, 3425.
4. M. Jones and K. R. Rettig, *J. Amer. Chem. Soc.*, 1965, **87**, 4013, 4015.
5. H. Nozaki, M. Yamabe, and R. Noyori, *Tetrahedron*, 1965, **21**, 1657.

6. S. D. McGregor and W. M. Jones, *J. Amer. Chem. Soc.*, 1968, **90**, 123.

7. W. J. Ball, S. R. Landor, and N. Punja, *J. Chem. Soc. (C)*, 1967, 194.

8. E. V. Dehmlow, *Chem. Ber.*, 1968, **101**, 410, 427.

9. R. W. Murray and M. L. Kaplan, *J. Amer. Chem. Soc.*, 1966, **88**, 3527.

10. For a recent review see G. Maier, *Angew. Chem. Internat. Edn.*, 1967, **6**, 402.

11. F. Nerdel, *et al.*, *Tetrahedron Lett.*, 1968, 1175.

12. C. W. Rees and C. E. Smithen, *J. Chem. Soc.*, 1964, 928.

13. A. Schönberg and R. von Ardenne, *Chem. Ber.*, 1968, **101**, 346.

14. H. Dworschak and F. Weygand, *Chem. Ber.*, 1968, **101**, 302.

15. A. Schönberg and B. König, *Chem. Ber.*, 1968, **101**, 725.

16. A. G. Cook and E. K. Fields, *J. Org. Chem.*, 1962, **27**, 3686.

17. E. P. Blanchard and H. E. Simmons, *J. Amer. Chem. Soc.*, 1964, **86**, 1337, 1347.

18. G. Köbrich, *Angew. Chem. Internat. Edn.*, 1967, **6**, 41.

19. G. L. Closs and R. A. Moss, *J. Amer. Chem. Soc.*, 1964, **86**, 4042; G. L. Closs and J. J. Coyle, *ibid.*, 1965, **87**, 4270.

Problems

5–1 Predict the products of the following reactions:

(a)

Photolysis of 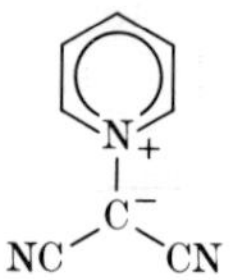   in benzene.

(J. Streith and J. M. Cassal, *Compt. Rend.*, 1967, **264**, 1307).

(b)

Thermolysis of 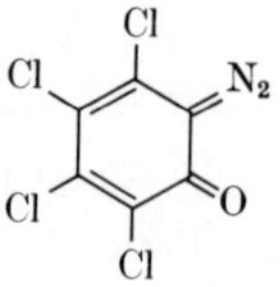   in styrene.

(R. Huisgen, H. König, G. Binsch, and H. J. Sturm, *Angew. Chem.*, 1961, **73**, 368).

5–2 The following useful reaction constructs a furan ring on ring A of a steroid. Suggest a mechanism for it.

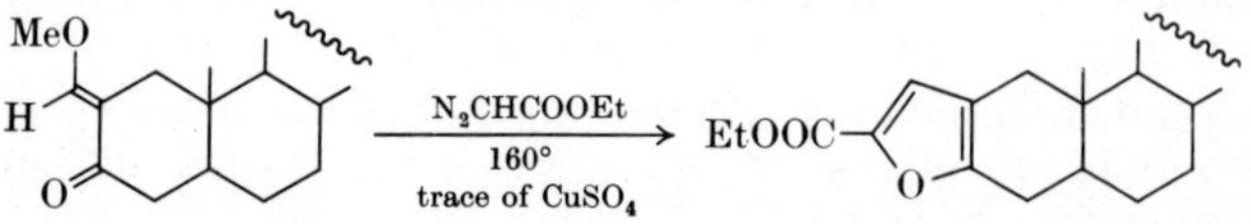

(D. L. Storm and T. A. Spencer, *Tetrahedron Lett.*, 1967, 1865).

5–3   When dimethyl diazomalonate, $N_2C(COOMe)_2$, is photolysed in *cis*- and *trans*-4-methyl-2-pentene, the corresponding cyclopropanes are formed in the following ratios:

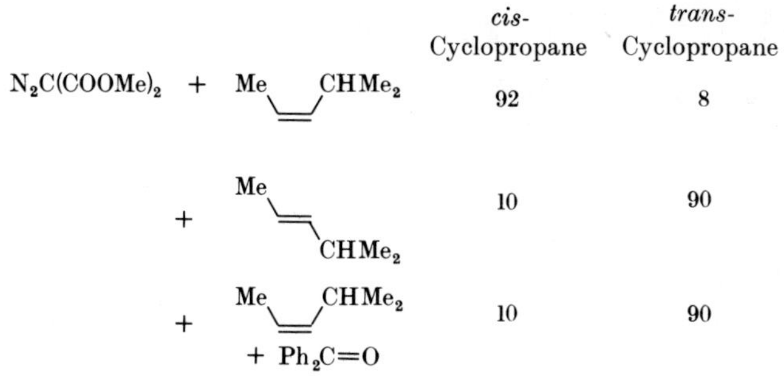

	cis- Cyclopropane	*trans*- Cyclopropane
$N_2C(COOMe)_2$ + Me⌣CHMe₂	92	8
+ Me⌣CHMe₂	10	90
+ Me⌣CHMe₂ + $Ph_2C=O$	10	90

Suggest an explanation.
(M. Jones, W. Ando, and A. Kulczycki, *Tetrahedron Lett.*, 1967, 1391).

5–4   Suggest possible mechanisms for:

(a)

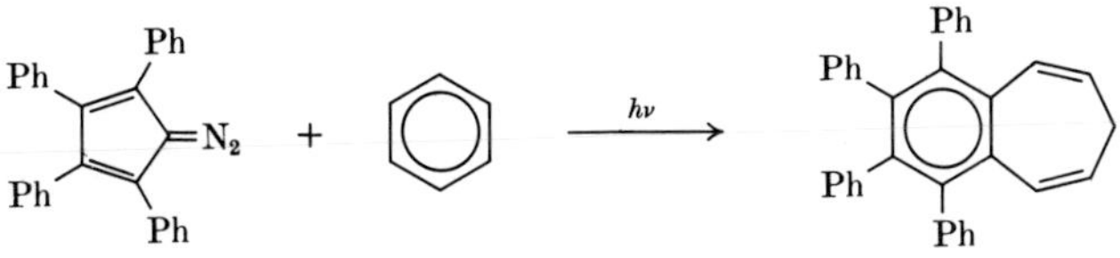

(H. Dürr and G. Scheppers, *Angew. Chem. Internat. Edn.*, 1968, **7**, 371).

(b)

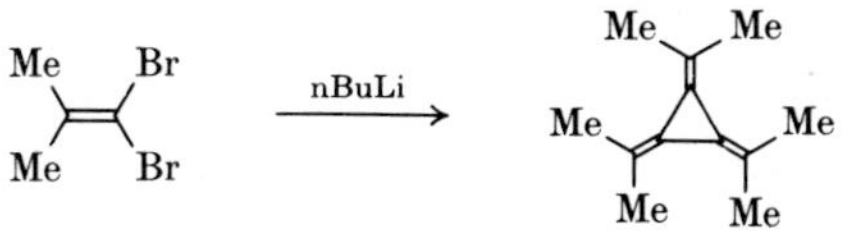

(G. Köbrich and H. Heinemann, *Angew. Chem. Internat. Edn.*, 1965, **4**, 594).

5–5   Compound A, when heated with dimethyl fumarate and a base, gives compound D in good yield. A is inert to base, but it rearranges and loses nitrogen when heated, to give an isolatable intermediate (B). With base, B reacts with dimethyl fumarate to give compound C, which rearranges with a trace of acid to its isomer (D). Suggest structures for B and C and account for the course of the reaction.

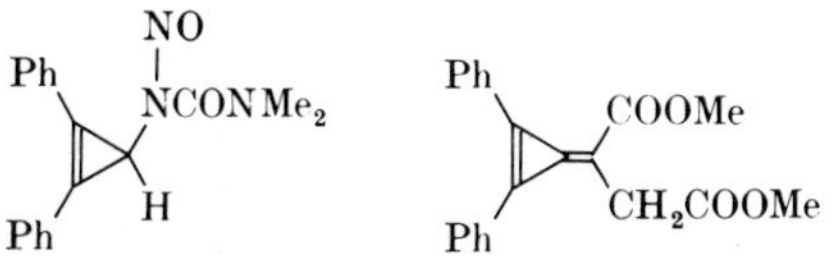

(W. M. Jones, M. E. Stowe, E. E. Wells, and E. W. Lester, *J. Amer. Chem. Soc.*, 1968, **90**, 1849).

6 Reactions of carbenes: insertions and other reactions

6–1 Insertion reactions

One of the most characteristic reactions of methylene is its insertion into C–H bonds:

$$\begin{array}{c} R \\ \diagdown \\ R-C-H \\ \diagup \\ R \end{array} + \; :CH_2 \longrightarrow \begin{array}{c} R \\ \diagdown \\ R-C-CH_3 \\ \diagup \\ R \end{array}$$

Many other carbenes can also insert into C–H bonds, either intramolecularly or intermolecularly. With methylene in the liquid phase, the insertion process is virtually indiscriminate, and all types of C–H bond in a substrate are attacked in a nearly statistical ratio. In the gas phase, there is some discrimination in favour of tertiary C–H bonds over secondary, and of secondary over primary. This selectivity is more pronounced with other carbenes: for example, ethoxycarbonylcarbene attacks tertiary C–H bonds about three times as readily as primary C–H bonds, and for diethoxycarbonyl-carbene the ratio is about ten to one. The selectivity of the carbene is probably related to the amount of excess of energy with which it is generated. It may also be connected with the spin state of the carbene: Skell's theory, that concerted addition to olefins is a characteristic of the singlet state, can be extended to insertions. Concerted and non-concerted mechanisms can be envisaged for the insertion reaction. The concerted process will involve a three-centre transition state:

$$\begin{array}{c} R \\ \diagdown \\ R-C-H \\ \diagup \\ R \end{array} + \; :CH_2 \longrightarrow \begin{array}{c} R \\ \diagdown \\ R-C\cdots\cdots H \\ \diagup \diagdown \vdots \diagup \\ R \quad C \\ \quad H_2 \end{array} \longrightarrow \begin{array}{c} R \\ \diagdown \\ R-C-CH_3 \\ \diagup \\ R \end{array}$$

The configuration of the substrate is retained in the product if this mechanism operates. An alternative is an abstraction–recombination process in which the carbene abstracts a hydrogen atom to give two radicals, which then recombine:

$$\underset{\underset{R}{|}}{\overset{\overset{R}{|}}{R-C-H}} + :CH_2 \longrightarrow \underset{\underset{R}{|}}{\overset{\overset{R}{|}}{R-C\cdot}} + \cdot CH_3 \longrightarrow \underset{\underset{R}{|}}{\overset{\overset{R}{|}}{R-C-CH_3}}$$

Since the mechanism involves radical intermediates, the configuration of the substrate will presumably be lost. We might also predict other products from the intermediate radicals, formed by rearrangement, abstraction, and dimerization, besides the recombination product, although the recombination could still be very efficient if the radicals were held close together, in a solvent 'cage', for example.

If the diradical mechanism operates, then the preferential insertion into tertiary C–H bonds can be explained by the greater stability of the intermediate tertiary radical: we would also expect that abstractions leading to especially stable radicals would be very favourable. It is more difficult to rationalize the preferential tertiary C–H insertion on the basis of the concerted mechanism. Doering has attempted to do this by proposing some polar character in the transition state:

$$\underset{\underset{R_2}{\overset{|}{C}}}{\overset{\diagdown}{\underset{\diagup}{-C}}\cdots\cdots H} \longleftrightarrow \underset{\underset{R_2}{\overset{|}{-C}}}{\overset{\diagdown}{\underset{\diagup}{-C}}^+} \quad \underset{\diagup}{\overset{H}{}}$$

The development of any carbonium ion character will be especially favoured with tertiary C–H bonds. Such a theory can explain the increasing selectivity of the carbenes $:CH_2$, $:CHCOOEt$, and $:C(COOEt)_2$, since this is the order in which they can increasingly stabilize the carbanion and therefore increase the polar nature of the transition state. There is some experimental evidence that can support the theory: bridgehead C–H bonds are unusually inert, and, for these, the development of carbonium ion character at the bridgehead is presumably much more difficult.

It is attractive to correlate the concerted mechanism with the reaction of carbenes in a singlet state, and the diradical mechanism with the reaction of triplet carbenes, by analogy with the olefin addition reactions. This is not a principle that can be applied to all insertions, however: the rare examples of insertion of dichlorocarbene into C–H bonds take place at allylic or benzylic carbons and lead to racemization at asymmetric centres. Thus for dichlorocarbene, which is generated as a singlet and has a singlet ground state, the mechanism seems to be of the abstraction–recombination type rather than a concerted one. A hydride abstraction is perhaps more likely than a radical abstraction here.

$$\geq\!C\!-\!H \ + \ :\!CCl_2 \ \longrightarrow \ \geq\!C^+ \ + \ \bar{C}HCl_2 \ \longrightarrow \ \geq\!C\!-\!CHCl_2$$

$$\begin{array}{c} Me \\ | \ * \\ Ph\!-\!C\!-\!H \\ | \\ Et \end{array} \ + \ :\!CCl_2 \ \longrightarrow \ \begin{array}{c} Me \\ | \\ Ph\!-\!C\!-\!CHCl_2 \\ | \\ Et \end{array} \ \ (inactive)$$

For other carbenes it is experimentally difficult to correlate spin state and the nature of the insertion reaction. Any attempt to remove triplet carbenes selectively is also likely to result in the removal of radical intermediates, since the scavenger will react with both. The stereospecificity of the reaction is probably the best test, and this has been correlated with spin state in some elegant experiments with nitrenes (Chapter 7). These results provide the best evidence that the Skell theory might apply to insertion reactions as well as to cyclo-additions.

Intramolecular insertions into C–H bonds are the preferred course of reaction of alkylcarbenes. These reactions are also selective, but the selectivity is governed more by steric factors (the size of the ring formed in an intramolecular cyclization, for example) than by a preference for a particular kind of C–H bond. The initial insertion product may also be formed with a large excess of energy, sufficient to cause it to isomerize before it is deactivated by collision, and this can also affect the product ratios. Only if several different sources of the carbene give the same product ratios can the effect of excess of energy be discounted. The major product of the intramolecular insertion is usually an olefin formed by insertion at the α-C–H bond; insertion at a β-C–H bond gives a cyclopropane as a second product:

$$\begin{array}{c} H \ \ R' \\ | \ \ \ | \\ R\!-\!C\!-\!C\!-\!\ddot{C}H \\ | \ \ \ | \\ R \ \ H \end{array} \ \longrightarrow \ \begin{array}{c} H \ \ R' \\ | \ \ \ | \\ R\!-\!C\!-\!C\!=\!CH_2 \\ | \\ R \end{array} \ + \ \begin{array}{c} R' \\ \triangle \\ R \end{array}$$

If more than one olefin can be formed by insertion into α-C–H bonds, then a tertiary or secondary C–H bond is attacked preferentially:

$$CH_3CH_2\ddot{C}CH_3 \longrightarrow \underset{\text{major product}}{CH_3CH\!=\!CHCH_3} + \underset{\text{minor product}}{CH_3CH_2CH\!=\!CH_2}$$

If the bivalent carbon is generated in a cyclic system, the products can include rings larger than cyclopropanes, formed by transannular insertions:

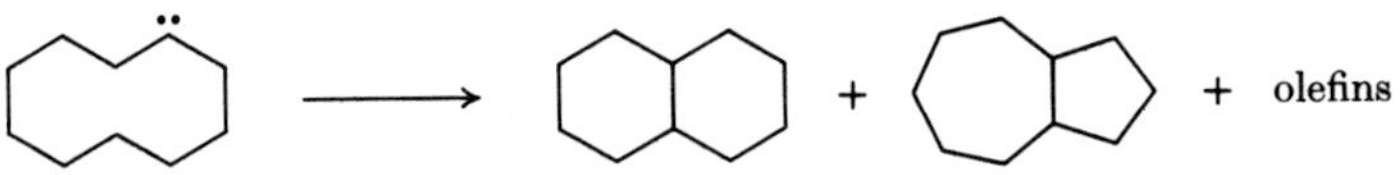

Alkoxycarbonylcarbenes and arylcarbenes can also insert intramolecularly into C–H bonds. In a hydrocarbon solvent there may be competing intermolecular insertion. Arylcarbenes will only insert into aromatic C–H bonds that are part of the same molecule:

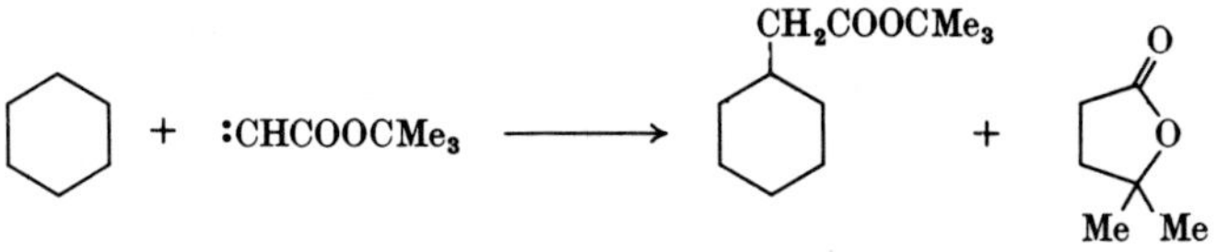

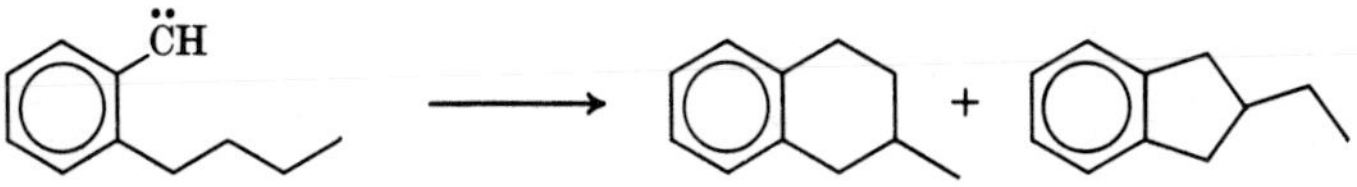

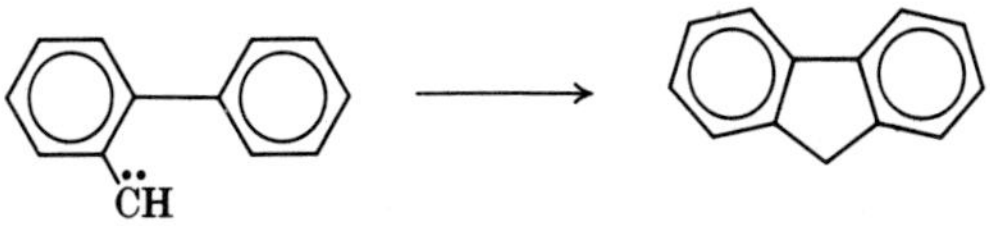

The insertion reaction is not confined to C–H bonds: it has been reported with many other types, including O–H, N–H, S–H, Si–H, Ge–H, C–Cl, C–Br, C–O, C–N, C–Si, and Sn–Sn bonds. The insertion of carbenes into C–Cl bonds is more rapid than into C–H bonds. It is believed to be the singlet carbenes that discriminate in favour of the C–Cl bonds. The mechanism is likely to be different from that of C–H insertion: with chlorine there is the possibility of an initial electrophilic attack by the carbene, followed by either a dissociation and recombination, or a rearrangement.

$$R_3C{-}Cl + :CR_2 \longrightarrow R_3C{-}Cl^{+}{-}\bar{C}R_2 \longrightarrow R_3C{-}CR_2Cl$$

With ethoxycarbonylcarbene, rearrangement products can be isolated from allyl halides:

$$MeCH{=}CHCH_2Cl + :CHCOOEt \rightarrow MeCH{=}CHCH_2CHClCOOEt + CH_2{=}CHCHMe\ CHClCOOEt$$

An initial electrophilic attack on oxygen is likely in the gas-phase insertion of methylene into the C–O bond of tetrahydrofuran[1] and in the insertion of ethoxycarbonylcarbene into 2-phenyloxetan:[2]

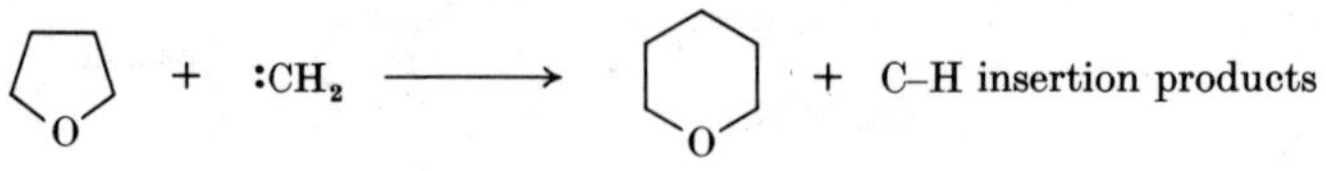

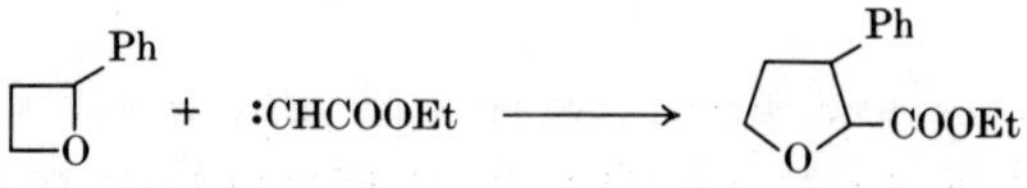

Similar ring-expansions are known with dichlorocarbene and cyclic silanes:[3]

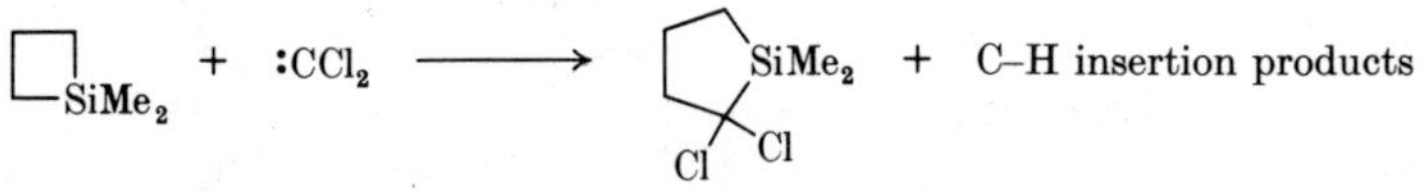

The insertions into O–H, N–H, S–H, and C–N bonds also involve an initial nucleophilic attack by the hetero atom.

The insertion of bromochlorocarbene into the tin–tin bond of hexamethylditin indicates the synthetic potential of the insertion reaction:[4]

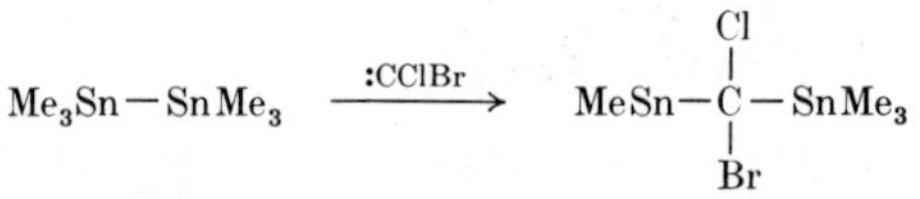

6–2   Skeletal rearrangements

Many of the rearrangements that have been ascribed to carbenes are 1,2-shifts of alkyl or aryl groups that can be represented by the general scheme:

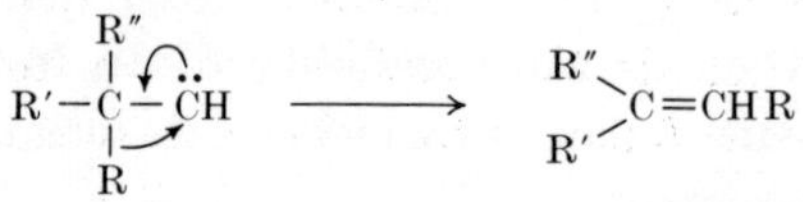

The evidence that these are rearrangements of the free carbenes and not their precursors (usually the diazoalkanes) is often flimsy. A concerted migration and expulsion of the leaving group is usually a valid alternative:

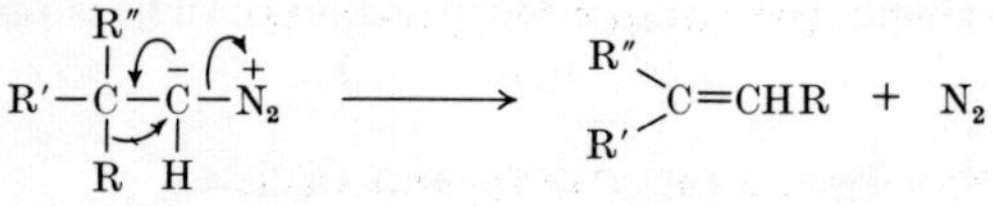

The most important of these is the Wolff rearrangement of diazo-ketones. This reaction is the key step in the Arndt–Eistert synthesis of aliphatic carboxylic acids:

$$RCOCHN_2 \longrightarrow RCH{=}C{=}O \xrightarrow{H_2O} RCH_2COOH$$

The evidence on the mechanism of the rearrangement does not allow a generalization in favour of either a carbene rearrangement or a concerted diazoketone rearrangement: it is likely that the nature of the substituents determines the timing of the steps. Optically active diazoketones rearrange with retention of configuration, but this does not distinguish between the two mechanisms.

$$\overset{*}{PhCH_2CHCOCHN_2} \xrightarrow[H_2O]{h\nu} \overset{*}{PhCH_2CHCH_2COOH}$$
$$\underset{Me}{\vert} \qquad\qquad\qquad \underset{Me}{\vert}$$

In the conditions of the Wolff rearrangement, other products that could be ascribed to oxocarbenes are rarely found. For example, the carbenes cannot be trapped by olefins, except in the presence of copper catalysts, when the rearrangement is suppressed. This can be interpreted as evidence against a carbene mechanism, and is in accord with the current view of the Curtius rearrangement of acylnitrenes (Chapter 7).

A rearrangement similar to this is also observed with diazo-sulphones, but, here, only about a tenth of the product is derived from the rearrangement :[5]

$$ArSO_2CHN_2 \longrightarrow ArCH{=}SO_2 + N_2$$
$$\downarrow MeOH \qquad\qquad \downarrow MeOH$$
$$ArSO_2CH_2OMe \quad ArCH_2SO_2Me$$

Again there is no conclusive evidence that free carbenes are inter-mediates. This is also true for a large number of rearrangements of diazoalkanes, in which a clear order of migratory aptitude is observed: hydrogen>aryl>alkyl.

$$\begin{array}{c} Me \\ \vert \\ Ph{-}C{-}CHN_2 \\ \vert \\ Me \end{array} \longrightarrow \begin{array}{c} Me \qquad H \\ \diagdown \qquad \diagup \\ C{=}C \\ \diagup \qquad \diagdown \\ Me \qquad Ph \end{array} + N_2$$

$$\begin{array}{c} Ph \\ \vert \\ H{-}C{-}CHN_2 \\ \vert \\ Ph \end{array} \longrightarrow \begin{array}{c} Ph \\ \diagdown \\ C{=}CH_2 \\ \diagup \\ Ph \end{array} + N_2$$

Some of these rearrangements are very useful synthetically, especially with diazoalkanes containing small rings, when an alkyl migration can lead to relief of the ring strain.

An example is the formation of cyclobutenes in good yield from α-cyclopropyl diazoalkanes. It is the less substituted bond of the cyclopropane that migrates preferentially:[6]

R
N₂
R
Me
R
R
Me

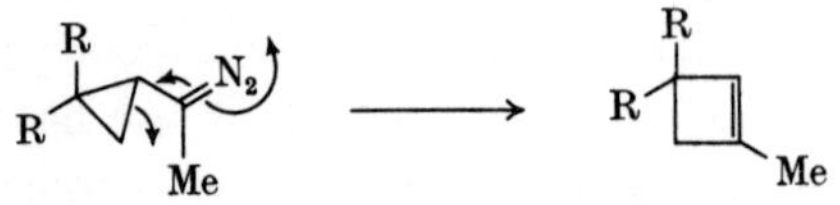

A similar rearrangement of cyclopropylidenes gives allenes. Though this rearrangement is usually ascribed to the singlet carbene,[7] it could again be a reaction of the precursor rather than of the free carbene. It is one of the most useful routes to allenes, because the carbene precursor can itself be made by adding dibromocarbene to an olefin:

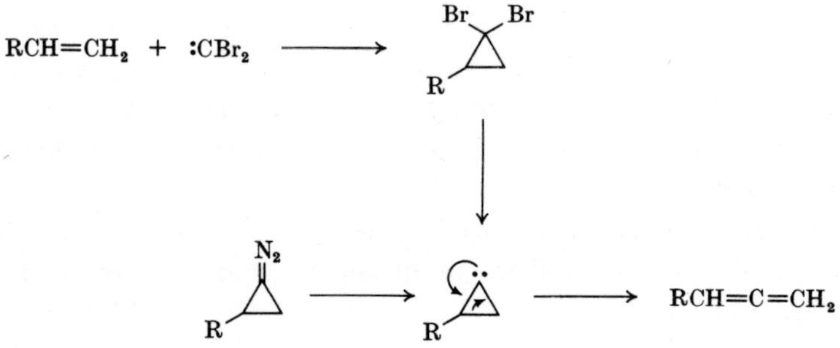

RCH=CH₂ + :CBr₂
Br
Br
R
N₂
R
R
RCH=C=CH₂

Similar reactions occur with diazoalkanes containing an extra nitrogen or sulphur dioxide molecule in the ring:

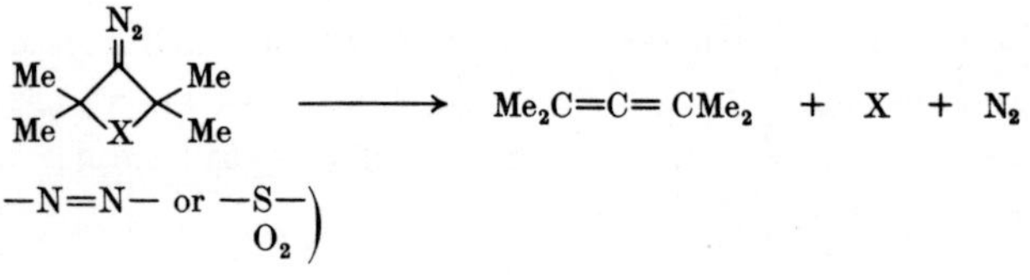

N₂
Me
Me
Me
Me
X
Me₂C=C=CMe₂ + X + N₂
(X = —N=N— or —S— O₂)

Attempts to detect the free carbene have failed.[8]

An analogous rearrangement of diazocyclobutanes can give methylenecyclopropanes.

Chloroalkoxycarbenes or dialkoxycarbenes are believed to rearrange by a 1,2-alkyl shift from oxygen to carbon, to give acid chlorides or esters. These rearrangements have been suggested to account for the formation of acid derivatives in reactions involving precursors of the carbenes:

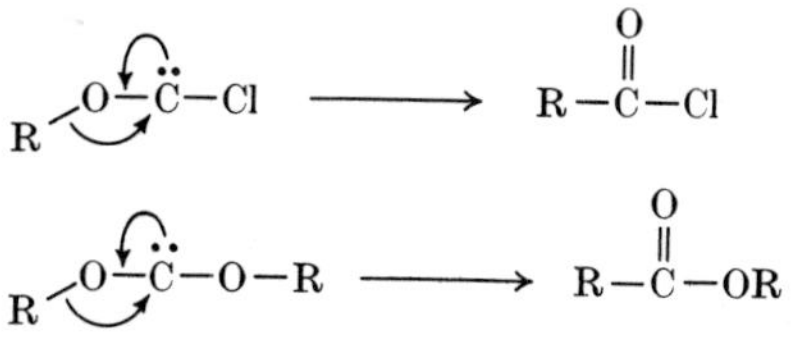

A different mechanism has to be proposed to account for some of the rearrangement products of highly strained diazoalkanes. In the thermolysis of the sodium salt of the toluene-*p*-sulphonylhydrazone [1], for example, hydrocarbons [2]–[5] are formed, and the suggested mechanism involves the diradical intermediate [6]. This may lose acetylene to give [2], or ring-close at C-1, C-3, or C-5 to give the bicyclic hydrocarbons [3], [4], or [5] respectively.[9]

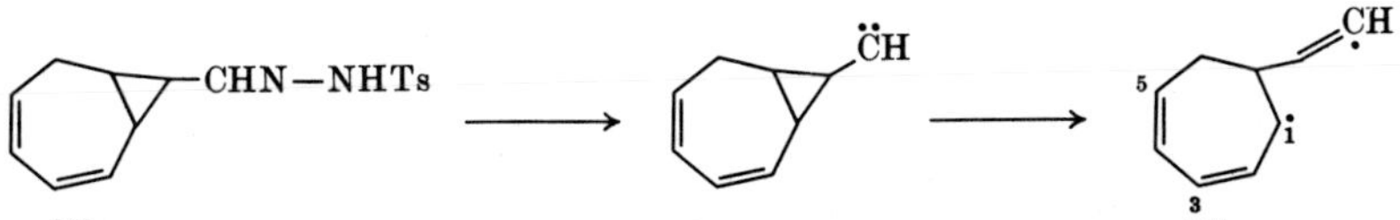

6–3   Reactions with nucleophiles

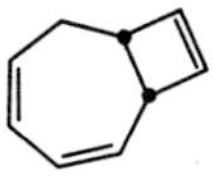

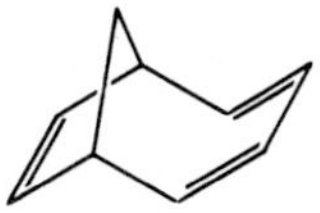

Carbenes can react with a wide variety of nucleophiles besides the nucleophilic olefins and others already described. Examples are known of reactions of carbenes with carbon, oxygen, nitrogen, sulphur, and phosphorus nucleophiles; these usually involve a relatively stable carbene like dichlorocarbene. For example, dichlorocarbene reacts with malonate anions to give the dichloromethyl derivatives:

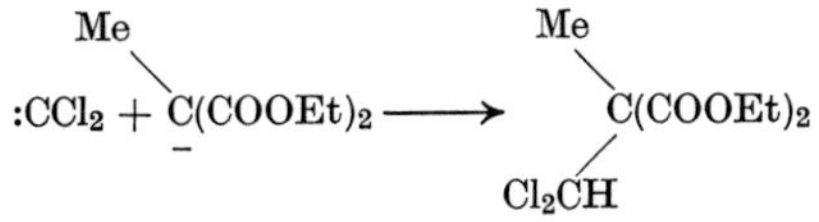

Several phosphorus ylids also react by nucleophilic attack of the carbanion on dichlorocarbene, to give dichloro-olefins.[10] Diazoalkanes react similarly at carbon with dichlorocarbene and with thiophenyl-carbene:[11]

$$Ph_2\bar{C}\!-\!\overset{+}{P}Ph_3 + :CCl_2 \longrightarrow Ph_2C\!=\!CCl_2 + PPh_3$$

$$Ph_2\bar{C}\!-\!\overset{+}{N}_2 + Cl\ddot{C}SPh \longrightarrow Ph_2C\!=\!CClSPh + N_2$$

The Reimer–Tiemann reaction[12] between phenoxide ions, chloroform, and base, to form salicylaldehyde, almost certainly involves an electrophilic attack by dichlorocarbene:

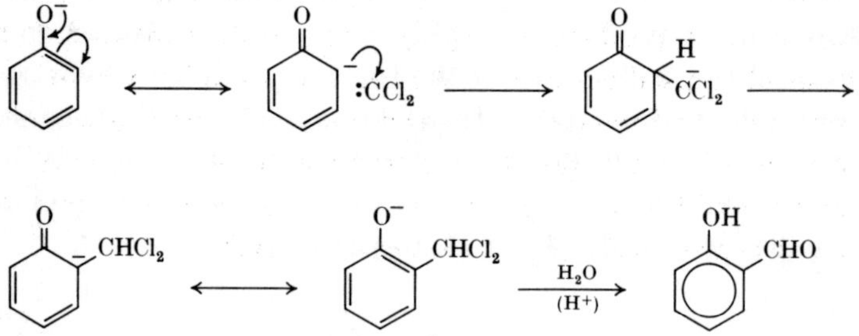

The products isolated from some 'abnormal' Reimer–Tiemann reactions lend support to this mechanism:

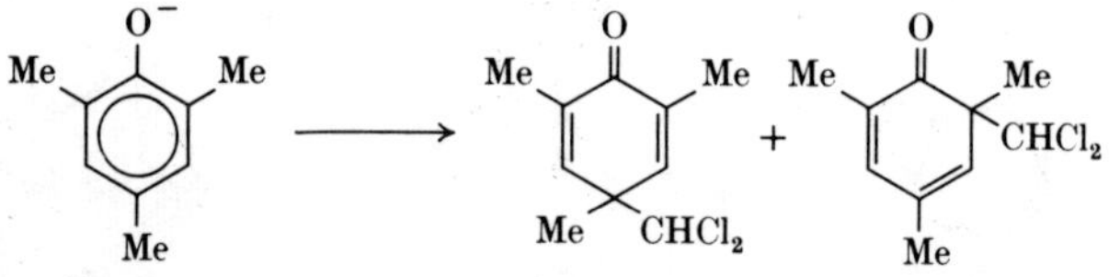

Several of the reactions of carbenes with oxygen-containing nucleophiles result in abstraction of oxygen by the carbenes. The preliminary step in all of these is presumably nucleophilic attack by the oxygen, as shown:

$$Me_2S\!=\!O + :CCl_2 \longrightarrow Me_2\overset{+}{S}\!-\!O\!-\!\overset{-}{C}Cl_2 \rightarrow Me_2S + COCl_2$$

$$F_3P\!=\!O + :CCl_2 \longrightarrow F_3P + COCl_2$$

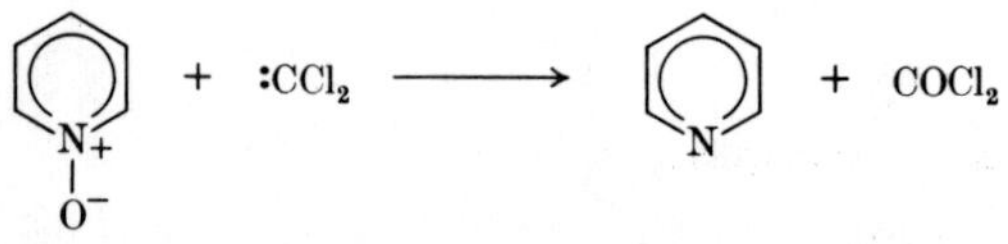

A few carbenes react with molecular oxygen to give ketones (diphenylcarbene gives benzophenone, for example), but this is almost certainly a radical reaction involving triplet carbenes only.

The ring-expansion reactions of cyclic ethers with carbenes probably involve a nucleophilic attack by oxygen (Section 6–1). With alcohols, the products are usually those of carbene insertion into the O–H bond. It is doubtful whether carbenes are involved in many of these, however; with diazoalkanes the alcohol probably protonates the precursor and the reaction goes through a carbonium ion. Hydroxide and alkoxide anions can react with carbenes, and these reactions are probably responsible for the by-products in the base-catalysed generation of dihalogenocarbenes from haloforms.

The reactions of carbenes with nitrogen, sulphur, and phosphorus nucleophiles are analogous. Sulphur has been used as a trap for nucleophilic carbenes.[13] With thiols and secondary amines, the products with carbenes are those of insertion into the S–H or N–H bonds. The well-known carbylamine test for primary amines with chloroform and base is probably a nucleophilic attack of the amine on dichlorocarbene followed by a double elimination of hydrogen chloride:

$$\text{RNH}_2 + \text{:CCl}_2 \longrightarrow R\overset{\overset{\displaystyle H}{|}}{\underset{\underset{\displaystyle H}{|}}{\overset{+}{N}}}\overset{-}{\text{CCl}_2} \longrightarrow R\overset{+}{-}\overset{-}{N}\equiv\overset{+}{C}$$

Sulphides[14] and tertiary amines[15] and phosphines give ylids as the primary products. The phosphorus ylids may be stable enough to undergo intermolecular reactions, but the sulphur ylids and nitrogen ylids are stabilized by intramolecular rearrangements:

$$\text{Ph}_3\text{P} + \text{:CCl}_2 \longrightarrow \text{Ph}_3\overset{+}{P}\overset{-}{-}\text{CCl}_2$$

$$\text{PhCH}_2\text{NMe}_2 + \text{:CCl}_2 \longrightarrow \text{PhCH}_2\overset{+}{\underset{\underset{\displaystyle \text{Me}_2}{|}}{N}}\overset{-}{-}\text{CCl}_2$$

$$\text{PhCH}_2\text{CCl}_2\text{NMe}_2 \longrightarrow \text{PhCH}_2\overset{\displaystyle \text{C}}{\underset{\underset{\displaystyle O}{\|}}{}}\text{NMe}_2$$

$$\text{MeSCH}_2\text{CH}{=}\text{CHMe} + \text{:CCl}_2 \longrightarrow \text{Me}\overset{+}{\underset{\underset{\displaystyle \text{CH}_2\text{CH}=\text{CHMe}}{|}}{S}}\overset{-}{-}\text{CCl}_2$$

$$\text{MeSCCl}_2\text{CH}_2\text{CH}{=}\text{CHMe}$$

The reactions between carbenes and diazoalkanes can go by a nucleophilic attack either by the carbon, or by the terminal nitrogen, on the carbene. The former is more common; in the latter, the product is an azine:

$$R_2C{=}N_2 + :CX_2 \longrightarrow R_2C{=}N{-}N{=}CX_2$$

Sometimes the azine is unstable and loses nitrogen in the reaction conditions. The product isolated is then the olefin $R_2C{=}CX_2$, which is the same as that which would be formed by direct nucleophilic attack by the carbon of the diazoalkane.

6–4 Dimerization

The dimerization of carbenes is a statistically unlikely process in most reactions, since they are such short-lived species, and their concentration will always be low. When formal carbene dimers are isolated, we can usually find a more likely mechanism for their formation. This is often an attack of the carbene on its precursor, as the reaction between a diazoalkane and a carbene in the previous section:

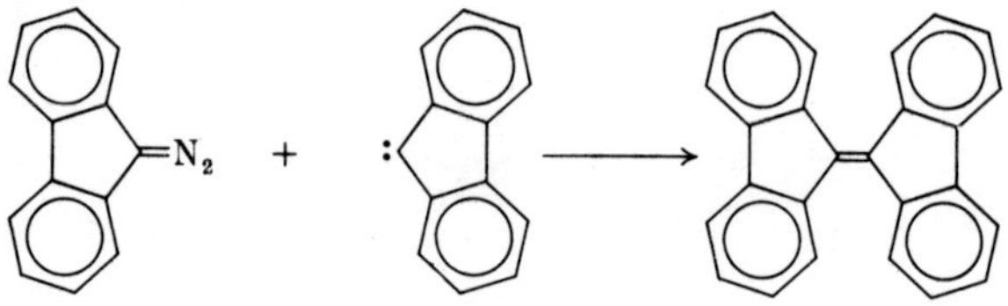

The formation of a dimer from the carbene produced by de-oxygenation of phthalic anhydride is similarly best explained by the attack of one carbene on an intermediate in the reaction, rather than by a simple dimerization:

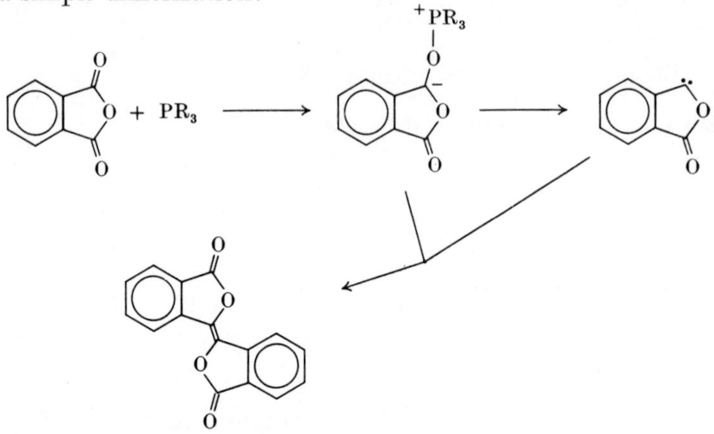

It has been suggested that some heavily hindered olefins can reversibly dissociate on heating into carbene monomers, the evidence being that carbene adducts can be isolated when the dimer is heated with a suitable trap. A more reasonable explanation is that the trap reacts first with the dimer, which then dissociates; for example:

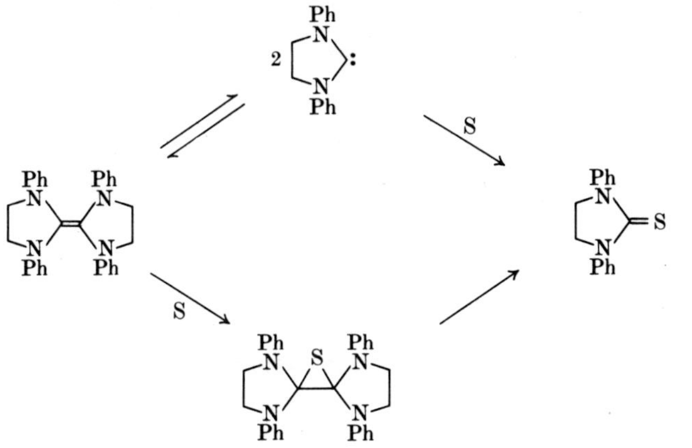

True carbene dimerizations are likely to be significant only in gas-phase reactions.

6–5   Carbenes in transition-metal complexes

Transition-metal complexes of reactive intermediates are theoretically interesting and potentially important in organic synthesis. The profound influence of copper salts and other transition-metal derivatives on the course of the decomposition of diazoalkanes indicates that carbene complexes may be intermediates. The potential application of such intermediates is illustrated by the asymmetric synthesis of cyclopropanes from methyl diazoacetate, styrene, and an optically active chelate as catalyst: the carbene may be complexed as a fifth, π-bonded, ligand to the copper:[16]

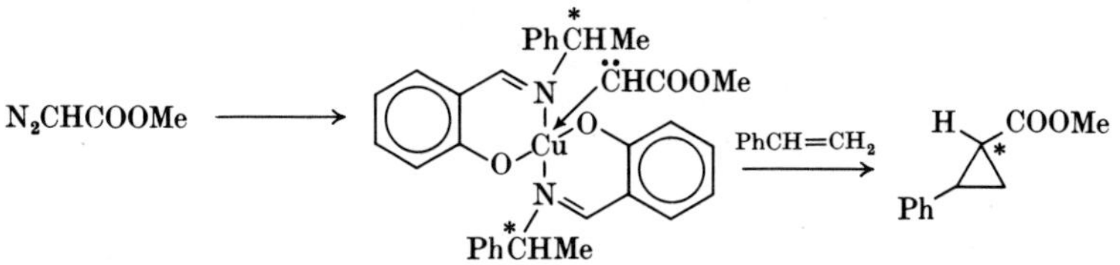

A similar π-bonded carbene complex has been postulated as the reactive form of bis(triphenylphosphine)chloromethyliridium carbonyl in solution:[17]

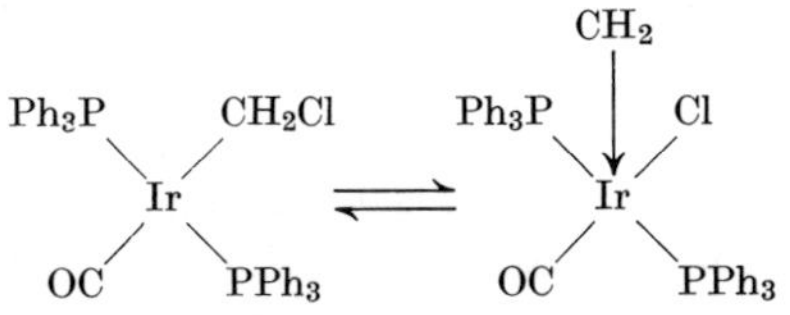

Methylene adducts and insertion products are obtained with styrene. The cyclopentadienyliron dicarbonyl methylene cation [7] has also been suggested as a reaction intermediate.[18] This methylene complex reacts stereospecifically with olefins to form cyclopropanes.

[7]

A series of methoxycarbene complexes of chromium, manganese, molybdenum, tungsten, and rhenium, of general structures [8] and [9], have also been described.[19]

$$\begin{array}{c} MeO \\ \diagdown \\ C-M(CO)_5 \\ \diagup \\ R \end{array}$$

(M = Cr, Mo, W)

[8]

$$\begin{array}{c} MeO \qquad CO \\ \diagdown \qquad \diagup \\ C-M-CO \\ \diagup \qquad \diagdown \\ R \qquad C_5H_5 \end{array}$$

(M = Mn, Re)

[9]

References

1. H. M. Frey and M. A. Voisey, *Chem. Comm.*, 1966, 454.
2. H. Nozaki, H. Takaya, and R. Noyori, *Tetrahedron*, 1966, **22**, 3393.
3. D. Seyferth, R. Damrauer, and S. S. Washburne, *J. Amer. Chem. Soc.*, 1967, **89**, 1538.
4. D. Seyferth and F. M. Armbrecht, *J. Amer. Chem. Soc.*, 1967, **89**, 2790.
5. R. J. Mulder, A. M. van Leusen, and J. Strating, *Tetrahedron Lett.*, 1967, 3057.
6. C. L. Bird, H. M. Frey, and I. D. R. Stevens, *Chem. Comm.*, 1967, 707.
7. W. T. Borden, *Tetrahedron Lett.*, 1967, 447.
8. R. Kalish and W. H. Pirkle, *J. Amer. Chem. Soc.*, 1967, **89**, 2781.
9. M. Jones and S. D. Reich, *J. Amer. Chem. Soc.*, 1967, **89**, 3935.
10. Y. Ito, M. Okano, and R. Oda, *Tetrahedron*, 1966, **22**, 2615.
11. H. Reimlinger, *Chem. and Ind.*, 1966, 1682.
12. H. Wynberg, *Chem. Rev.*, 1960, **60**, 169.
13. E. Winterfeldt and G. Giesler, *Angew. Chem. Internat. Edn.*, 1966, **5**, 579.
14. W. E. Parham and S. H. Groen, *J. Org. Chem.*, 1966, **31**, 1694.
15. M. Saunders and R. W. Murray, *Tetrahedron*, 1960, **11**, 1.
16. H. Nozaki, S. Moriuti, H. Takaya, and R. Noyori, *Tetrahedron*, 1968, **24**, 3655.
17. F. D. Mango and I. Dvoretzky, *J. Amer. Chem. Soc.*, 1966, **88**, 1654.
18. P. W. Jolly and R. Pettit, *J. Amer. Chem. Soc.*, 1966, **88**, 5044.
19. E. O. Fischer and A. Riedel, *Chem. Ber.*, 1968, **101**, 156.

Problems

6–1 Consider all the formally possible products of the photolysis of ArSO$_2$CCOAr in ethanol, and suggest which are more likely to be formed
$$\overset{\|}{N_2}$$
and why.

(A. M. van Leusen, P. M. Smid, and J. Strating, *Tetrahedron Lett.*, 1967, 1165).

6–2 Suggest possible explanations for the following:

(a)

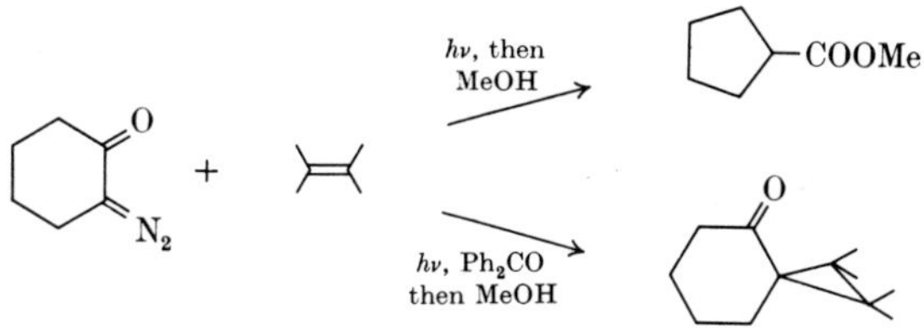

(M. Jones and W. Ando, *J. Amer. Chem. Soc.*, 1968, **90**, 2200).

(b)

$$N_2CHCOOEt \xrightarrow[Me_2CHOH]{h\nu} EtOCH_2COOCHMe_2 + Me_2CHOCH_2COOEt$$

(O. P. Strausz, T. DoMinh, and H. E. Gunning, *J. Amer. Chem. Soc.*, 1968, **90**, 1660).

(c)

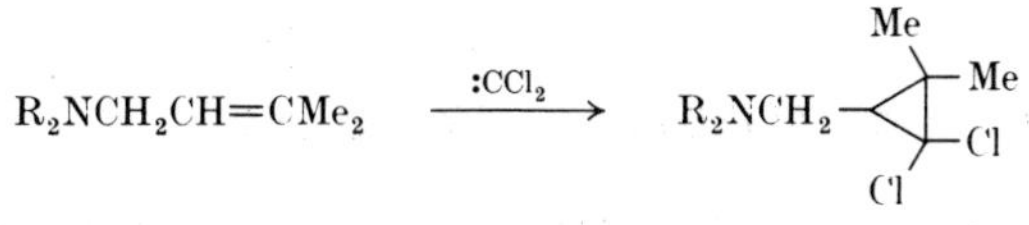

but RSCH$_2$CH=CR$_2$ gives no cyclopropane in similar conditions; the major product is RSCCl$_2$CR$_2$CH=CH$_2$.
(W. E. Parham and J. R. Potoski, *J. Org. Chem.*, 1967, **32**, 275).

(d) Photolysis of diphenyldiazomethane gives mainly benzophenone azine (A), but photolysis of dimesityldiazomethane gives exclusively tetramesitylethylene (B).

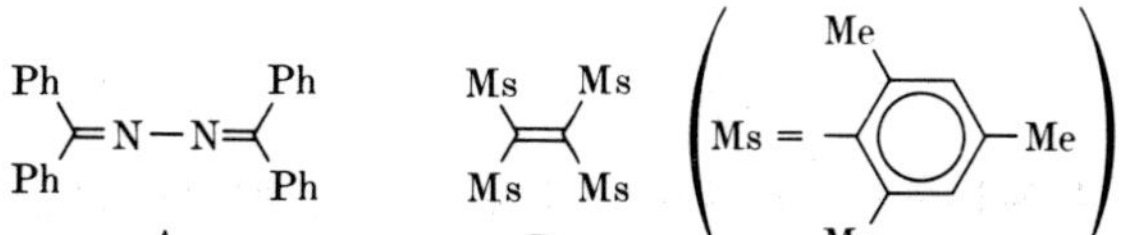

(H. E. Zimmerman and D. H. Paskovich, *J. Amer. Chem. Soc.*, 1964, **86**, 2149).

6–3   Suggest possible mechanisms for the following:

(a)

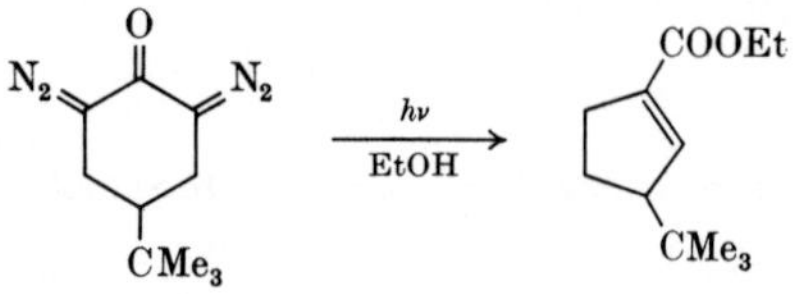

(R. Tasovac, M. Stefanović, and A. Stojiljković, *Tetrahedron Lett.*, 1967, 2729).

(b)

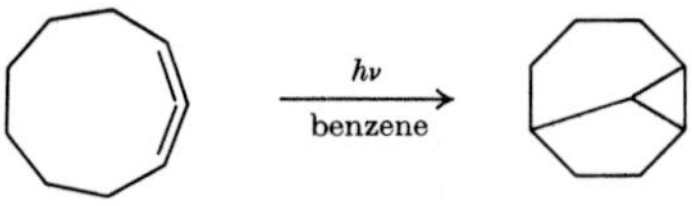

(H. R. Ward and E. Karafiath, *J. Amer. Chem. Soc.*, 1968, **90**, 2193).

(c) Both 2- and 4-methylpyridine are reported to be converted by chloroform and concentrated alkali into phenyl isocyanide.

(J. Ploquin, *Bull. Soc. Chim. France*, 1947, 901).

(d)

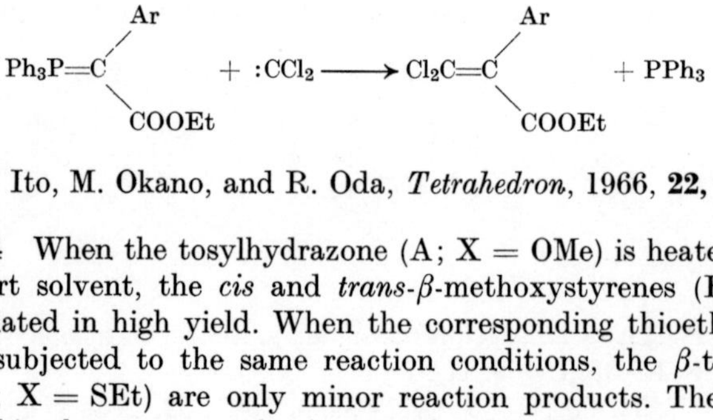

(Y. Ito, M. Okano, and R. Oda, *Tetrahedron*, 1966, **22**, 2615).

6–4   When the tosylhydrazone (A; X = OMe) is heated with base in an inert solvent, the *cis* and *trans*-β-methoxystyrenes (B; X = OMe) are isolated in high yield. When the corresponding thioether (A; X = SEt) is subjected to the same reaction conditions, the β-thioethoxystyrenes (B; X = SEt) are only minor reaction products. The major product is α-thioethoxystyrene (C). Account for this difference.

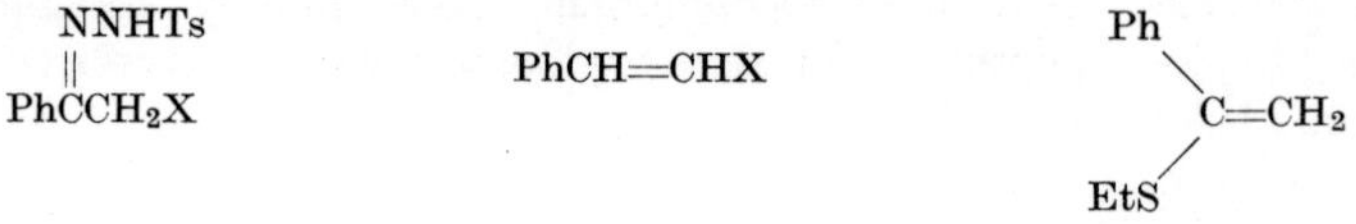

(J. H. Robson and H. Schechter, *J. Amer. Chem. Soc.*, 1967, **89**, 7112).

Reactions
of nitrenes

7–1 Introduction

The broad patterns of reactivity that have been established in carbene chemistry are also emerging with nitrenes. All the characteristic carbene reactions (additions to olefins and to aromatic compounds, insertions, abstraction of hydrogen, intramolecular 1,2-shifts, and others) find their parallels in nitrene chemistry. So far, the experimental data for nitrene reactions are much more fragmentary than for carbenes, and often there is no rigorous proof that nitrenes are intermediates. It is usually possible to account for the reaction products by an alternative mechanism that does not involve nitrenes.

The most extensive work has been done with ethoxycarbonylnitrene, :ṄCOOEt, mainly by Lwowski and his co-workers.[1] The same products have been obtained from very different precursors of the nitrene, which is thus the only reasonable common intermediate. The chemistry of this nitrene is broadly similar to that of the corresponding carbene, :CHCOOEt. We can reasonably expect this similarity to extend to other nitrenes and carbenes.

7–2 Cycloaddition reactions

Some nitrenes can be trapped with olefins to give aziridines, a few apparently cannot, but, with most of the nitrenes postulated as intermediates, no attempt has so far been made to trap them. The reason for this is that, until recently, the only general sources of nitrenes were the corresponding azides, and with azides there is a fundamental difficulty: they can react with olefins to give triazolines, which may then lose nitrogen to form aziridines:

It is therefore difficult to distinguish between a triazoline and a nitrene mechanism if aziridines are produced. If the azide is decomposed thermally, then the formation and decomposition of a triazoline might be indicated by an enhanced rate of nitrogen evolution. This is observed with phenyl azide and indene or styrene, for example. If the azide is photolysed, nothing can be deduced about the mechanism from the rate of gas evolution, which depends on the light flux and the quantum yield.

The difficulty is overcome if the same aziridines, in the same proportions, can be obtained from the azide and from an entirely independent nitrene precursor. Ethoxycarbonylnitrene gives the same products with olefins whether it is formed by thermolysis or photolysis of the azide, or by an α-elimination route. The results agree very well with the Skell hypothesis of stereospecific and non-stereospecific addition by singlets and triplets, respectively. (See Chapter 5 for a similar discussion of carbene additions to olefins.)

With cis-4-methyl-2-pentene, for example, a mixture of two aziridines is obtained, the cis-aziridine [1] having the stereochemistry of the olefin retained, and the trans-aziridine [2] having lost it:

In the pure olefin as solvent, the product is largely the cis-aziridine, indicating that the reaction is mainly stereospecific, but, as the olefin is increasingly diluted with an inert solvent, the proportion of the trans-aziridine [2] goes up. The ground state of the nitrene is a triplet, but the nitrene is formed in a singlet state; at high olefin concentrations the singlet nitrene adds stereospecifically to give [1], but with increasing dilution there is more chance of the nitrene being converted by collisions into the ground-state triplet, which then adds non-stereospecifically to give [1] or [2]:

The triplet nitrene can be selectively removed by adding a good radical trap, α-methylstyrene. This olefin reacts very much faster with the triplet than an ordinary nucleophilic olefin, but it only reacts very slowly with the singlet:

The result of adding α-methylstyrene is therefore to increase the ratio of [1] to [2], since only the singlet nitrene is able to react with the 4-methyl-2-pentene.

Ethoxycarbonylnitrene adds to other nucleophilic olefins, such as cyclohexene, *cis*- and *trans*-butene, and dihydropyran, to give good yields of the corresponding aziridines. The nitrene, like the corresponding carbene, can also add as a 1,3-dipole. It reacts in this way with nitriles and with arylacetylenes:

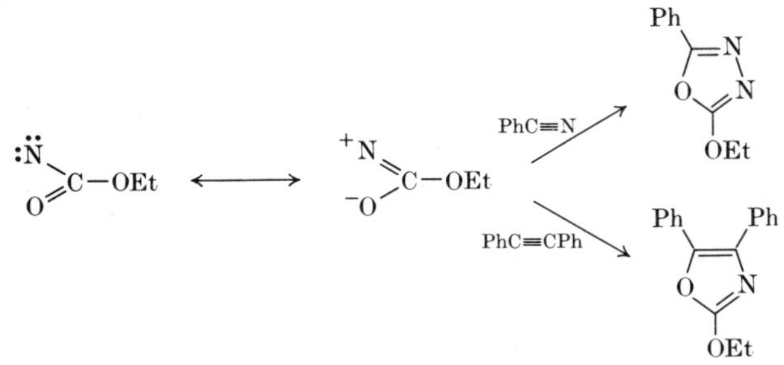

The available evidence on the addition of other nitrenes to olefins seems to indicate that the nitrene must have some enhanced stability before it can be trapped. The symmetrical cyanonitrene, for example, reacts readily with cyclo-octatetraene, giving the aziridine [3] and the pyrroline [4] among the products:[2]

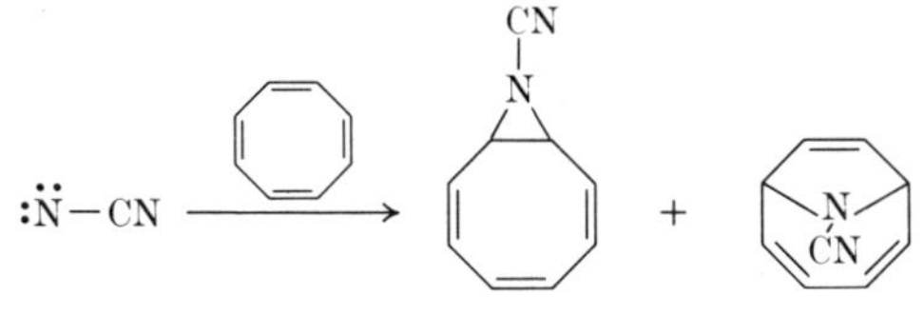

The adduct [4] is a rare example of a product of 1,4-addition of a nitrene to a conjugated polyene. It is believed to be formed from the triplet nitrene. By contrast, ethoxycarbonylnitrene seems to add exclusively 1,2 to conjugated dienes.[3]

The other nitrenes that are known to add to olefins are the resonance-stabilized aminonitrenes.[4] Even these nitrenes will only add if they cannot react readily by some intramolecular pathway, such as rearrangement or loss of nitrogen. The nitrenes below are examples of several groups of intermediates that can be generated by oxidation of the corresponding amines, and trapped with olefins. With some, such as phthalimidonitrene [5], the same intermediates can be generated by photolysis of the corresponding sulphoximines.

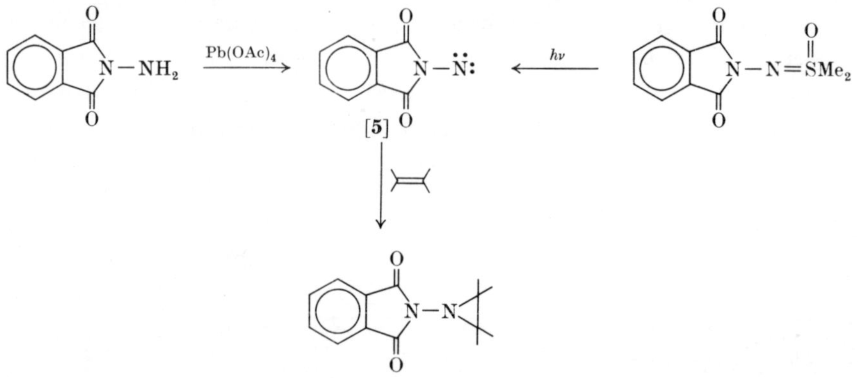

Similar:

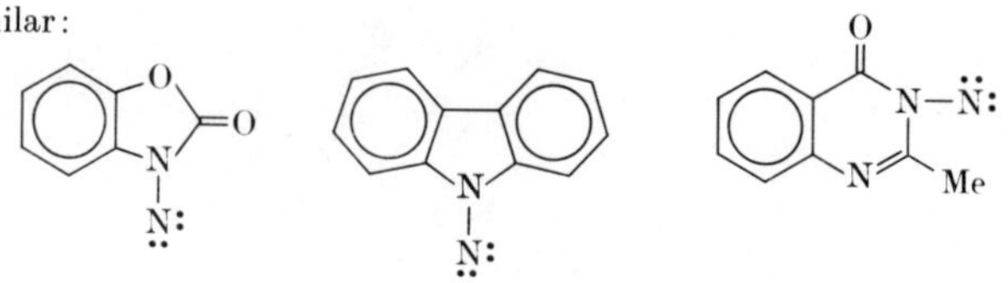

These nitrenes add stereospecifically to nucleophilic olefins like *cis*- and *trans*-butene. They can also be trapped by a wide range of other olefins, including styrene and conjugated dienes. The nitrenes can also add to electrophilic olefins, such as unsaturated ketones and esters: methyl acrylate, ethyl acrylate, mesityl oxide, diethyl fumarate, and several others. In this they resemble nucleophilic carbenes (Chapter 5).

Aziridines are also found, with a multiplicity of other products, in the copper-catalysed decomposition of sulphonyl azides in cyclohexene:[5]

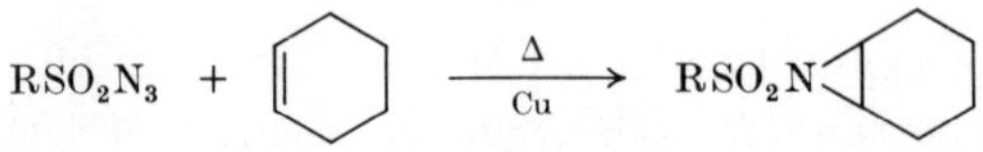

The addition probably involves a copper–nitrene complex. The copper does not seem to be so effective in suppressing side reactions as it is in the analogous carbene additions; several of the products are those of insertion.

Simple alkylnitrenes probably rearrange too rapidly to be trapped by olefins. Only intramolecular double bonds are likely to be capable of trapping these nitrenes, as, for example, the nitrene that may be formed by oxidizing 4-(β-aminoethyl)cyclohexene[6] (Chapter 3).

Just as energetic carbenes can attack benzene rings to give ring-expanded products, so certain nitrenes can give azepines with aromatic compounds. Ethoxycarbonylnitrene, even when it is generated by the α-elimination route, gives the corresponding azepine from benzene in good yield:

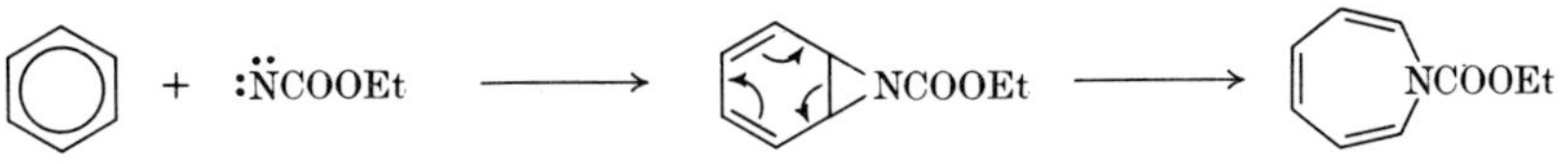

This is the best way of making the interesting azepine ring system. A similar reaction is believed to occur with sulphonylnitrenes, but the initial adduct rearranges to a sulphonylaniline derivative instead of ring-expanding, possibly because of the greater stability of the developing sulphonamide anion:

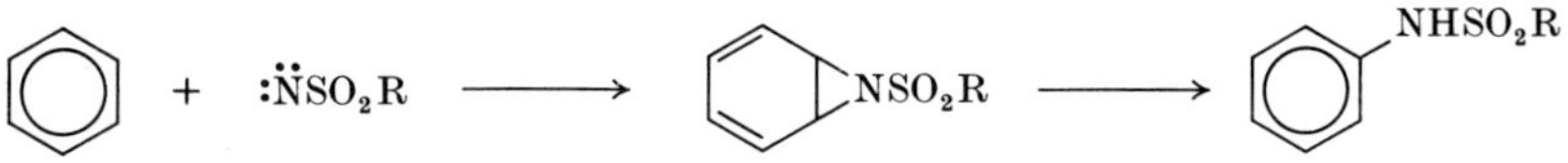

An interesting reaction, which may be of this type, takes place when phenyl azide is photolysed in amine solvents.[7] Azepine derivatives are formed: the most reasonable mechanism involves an intramolecular attack of the nitrene on the phenyl ring:

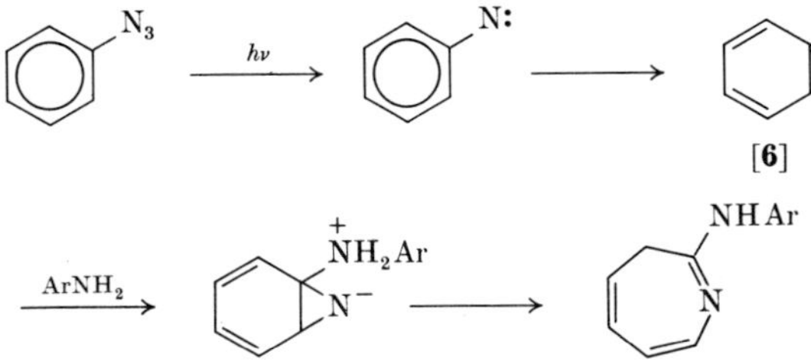

The course of the reaction is very dependent on the conditions: for example, it goes by this route in aliphatic and aromatic amines and with hydrogen sulphide, but not in hydrocarbon solvents or with ammonia. There is some analogy for this strained azirine intermediate [6], since azirines are formed, often in high yield, in the photolysis of vinyl azides:[8]

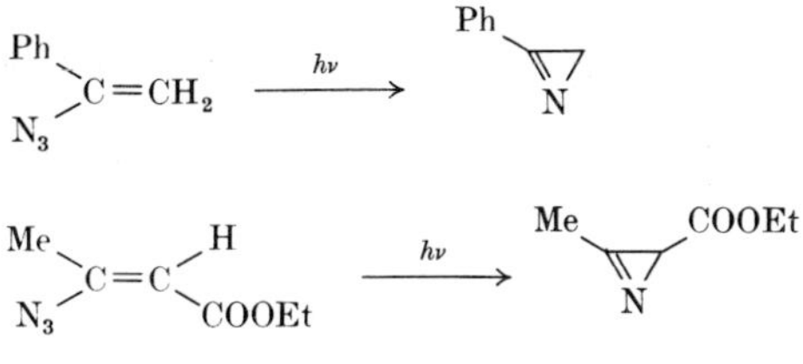

The intermediacy of nitrenes in these reactions has not been established, however; the ring-closure may be concurrent with the decomposition of the azide.

7–3 Insertion reactions

Just as with carbenes, 'insertion' in nitrene chemistry describes the overall reaction rather than the detailed mechanism, though it is sometimes used to imply a concerted process. If the Skell hypothesis applies to the insertion of nitrenes into C–H bonds, then singlet nitrenes should insert in a concerted manner, and the stereochemistry of the substrate should be retained in the product:

$$R{-}\ddot{N}{\uparrow\downarrow} + {-}\overset{\diagdown}{\underset{\diagup}{C}}{-}H \longrightarrow {-}\overset{\diagdown}{\underset{\diagup}{C}}{\cdots}H \longrightarrow {-}\overset{\diagdown}{\underset{\diagup}{C}}{-}NHR$$

Triplet nitrenes could give the same products, but by a two-step process: abstraction of a hydrogen by the nitrene, followed by the combination of the two radicals. This reaction need not be stereospecific if the carbon radical can become planar before it combines with the nitrogen radical (as for the corresponding carbene reactions: Chapter 6):

$$R{-}\ddot{N}{\uparrow\uparrow} + {-}\overset{\diagdown}{\underset{\diagup}{C}}{-}H \longrightarrow {-}\overset{\diagdown}{\underset{\diagup}{C}}{\cdot} + RN{\cdot} \longrightarrow {-}\overset{\diagdown}{\underset{\diagup}{C}}{-}NHR$$

Some elegant work with different nitrenes bears out these conclusions.

The insertion of cyanonitrene into the tertiary C–H bonds of *cis*- and *trans*-1,2-dimethylcyclohexane is stereospecific when the hydrocarbons are used as the solvent:[9]

In methylene chloride solvent, the reaction becomes less stereospecific, and in methylene bromide, a good solvent for deactivating the nitrenes by collisions, the stereospecificity is lost completely. The nitrene, first generated as an excited singlet, inserts in a concerted fashion into the C–H bonds, but in the presence of methylene chloride or methylene bromide the nitrene is converted into the ground-state triplet by collisions with solvent molecules before it can insert into the hydrocarbon.

With ethoxycarbonylnitrene, the insertion into tertiary C–H bonds is stereospecific at *all* concentrations of the hydrocarbon in the solution.[1] From the results of the addition of this nitrene to olefins, we know that the proportion of triplet to singlet nitrenes must be high, especially in dilute solutions, so the triplet presumably does not insert at all into the tertiary C–H bonds. The yield of insertion product falls off rapidly as the hydrocarbon concentration in the solution is decreased.

If the tertiary C–H bond is at an asymmetric carbon atom, then the optical activity should be retained in a concerted insertion. This retention of optical activity has been observed in a few intramolecular insertion reactions,[10] for example:

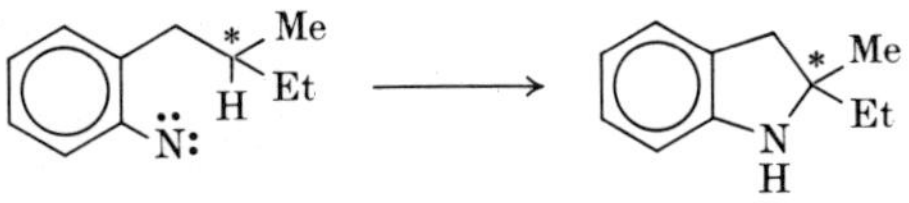

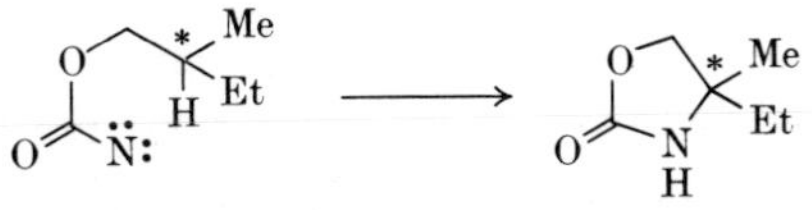

Cyclizations of this sort take place only when more favourable reactions, such as 1,2-hydrogen shifts, are precluded, and when an unstrained ring is produced.

The reaction goes best with arylnitrenes: thus carbazole can be produced from *o*-azidobiphenyl in good yield:

This reaction fails when there is a neighbouring group in the molecule, such as a nitro or carbonyl, which will trap the nitrene preferentially:

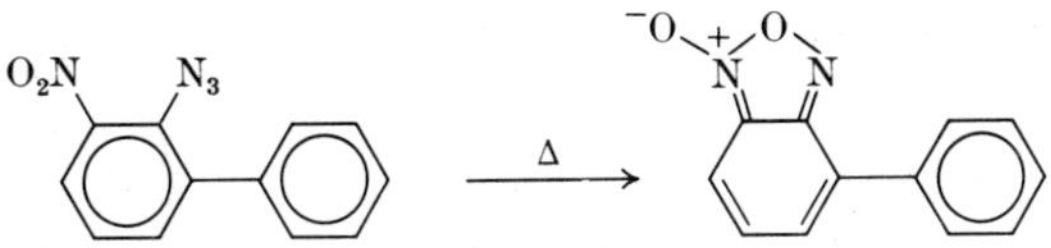

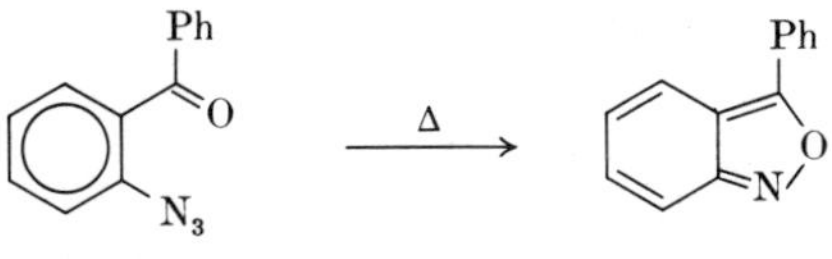

Intermolecular insertions by nitrenes, though common, are probably of less synthetic value than their intramolecular counterparts, because they are not very discriminating. With C–H bonds, there is preferential insertion into tertiary C–H, followed by secondary and primary C–H bonds: the relative reactivities with ethoxy-carbonylnitrene are about 30:10:1. Even the most favourable insertion process is only about a fifth as likely as addition to the double bond, when the solvent is an olefin.

The abstraction of hydrogen from the solvent is a common side reaction with arylnitrenes and sulphonylnitrenes, the products being aromatic primary amines and sulphonamides respectively:

$$Ar\ddot{N}: \longrightarrow ArNH_2$$

$$RSO_2\ddot{N}: \longrightarrow RSO_2NH_2$$

The obvious mechanism, stepwise abstraction of hydrogen by the triplet nitrene, has been deemed unlikely for sulphonylnitrenes in aromatic solvents because of the absence of the expected free-radical products.[11] The mechanism therefore remains obscure.

The insertion reactions of nitrenes are not confined to C–H bonds. For example, ethoxycarbonylnitrene reacts with alcohols and amines to give the corresponding hydroxylamine and hydrazine derivatives:

$$:\ddot{N}COOEt \quad \overset{ROH}{\underset{R_2NH}{\Big\langle}} \quad \begin{array}{l} RONHCOOEt \\[1em] R_2NNHCOOEt \end{array}$$

Even these reactions are not general, however; arylnitrenes give azepine derivatives in amine solvents (Section 7–2).

7–4   1,2-Shifts

Several reactions of nitrenes can be represented by the general scheme:

$$\underset{R}{\overset{\displaystyle >C}{\diagdown}}\,\ddot{N}: \longrightarrow \quad >C=NR$$

These reactions are, in effect, a class of intramolecular insertions. They are especially common when R = H, and result in the formation of an imine, which may hydrolyse to the corresponding aldehyde or ketone in the conditions of the reaction:

$$\overset{H}{\underset{}{>\!\!C}}-\ddot{N}: \longrightarrow \quad >C=NH \quad \overset{H_2O}{\longrightarrow} \quad >C=O$$

Imines or the corresponding carbonyl compounds are often the major products of the decomposition of alkyl azides. We should emphasize that, although the nitrene mechanism is a possible way of accounting for these products, there is as yet no direct evidence for nitrene intermediates, and a concerted rearrangement and loss of nitrogen is just as likely.[12]

$$R[CH_2]_4N_3 \xrightarrow{h\nu} R[CH_2]_3CH{=}NH$$

$$Me[CH_2]_2N_3 \longrightarrow MeCH_2CH{=}NH \longrightarrow MeCH_2CHO$$

$$MeN_3 \longrightarrow CH_2{=}NH \longrightarrow \text{decomposition products}$$

$$\overset{\overset{\displaystyle H}{|}}{RCOCN_3} \xrightarrow{\Delta} \overset{\overset{\displaystyle}{}}{RCOC}{=}NH$$
$$\underset{\underset{\displaystyle R'}{|}}{} \qquad\qquad \underset{\underset{\displaystyle R'}{|}}{}$$

(R = phenyl, o-hydroxyphenyl, p-hydroxyphenyl, ethoxy)

If the carbon atom next to the electron-deficient nitrogen does not carry a hydrogen atom, then other groups can migrate. Aryl groups migrate in preference to alkyl groups:

$$\overset{\overset{\displaystyle Me}{|}}{PhCOCN_3} \xrightarrow{\Delta} \overset{\overset{\displaystyle Me}{|}}{PhCOC}{=}NPh$$
$$\underset{\underset{\displaystyle Ph}{|}}{}$$

$$Ar_3CN_3 \longrightarrow Ar_2C{=}NAr$$

$$CF_3CHFCF_2N_3 \longrightarrow CF_2{=}NCHFCF_3$$

The Curtius rearrangement of carbonyl azides, once thought to be a nitrene reaction of this type, probably does not involve free nitrenes at all (see Chapter 3). Rearrangements that are formally similar to the Curtius rearrangement occur as minor reaction paths in the photolysis of vinyl azides.[8] These may also be concerted reactions rather than reactions of free nitrenes.

$$\overset{\displaystyle R}{\underset{\displaystyle O}{\overset{\displaystyle \diagdown}{\underset{\displaystyle \|}{C}}}} \overset{\displaystyle \bar{N}{-}N{=}\overset{+}{N}}{} \xrightarrow{-N_2} R{-}\bar{N}{-}\overset{+}{C}{=}O \longrightarrow R{-}N{=}C{=}O$$
$$\text{(Curtius rearrangement)}$$

$$\overset{\displaystyle R}{\underset{\displaystyle CH_2}{\overset{\displaystyle \diagdown}{\underset{\displaystyle \|}{C}}}} \overset{\displaystyle \bar{N}{-}N{=}\overset{+}{N}}{} \xrightarrow{-N_2} R{-}N{=}C{=}CH_2$$

Similar rearrangements can probably take place with some amino-nitrenes. A 1,2-aryl shift is observed in some oxidations of N-amino compounds.[4,13] The mechanism shown is a possible one, but it has not been proved:

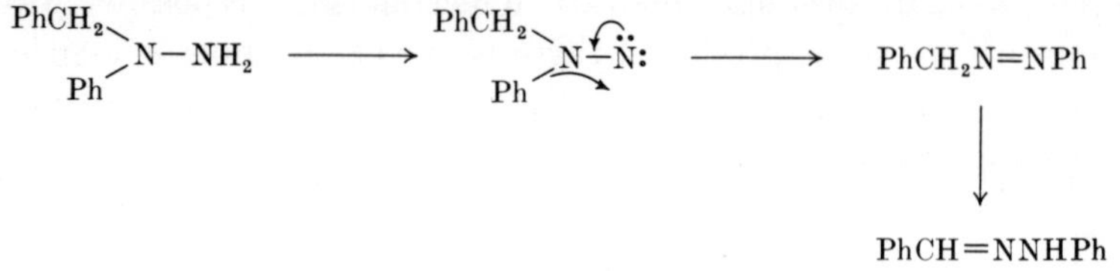

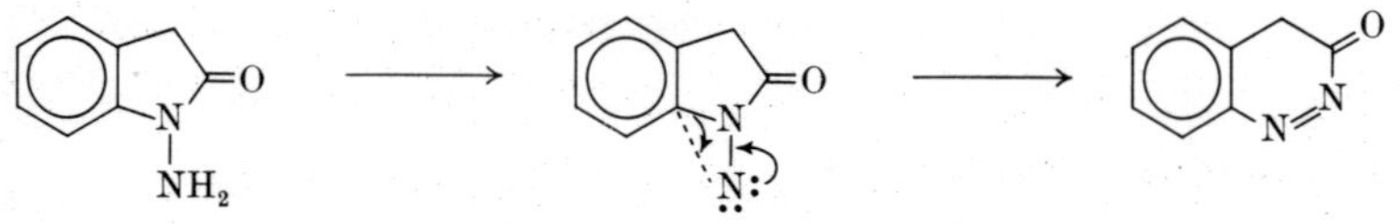

7–5   Dimerization

In several reactions that could involve nitrenes, the products include azo compounds, which are the formal nitrene dimers:

$$2RN: \longrightarrow RN=NR$$

The dimerization of nitrenes in solution is not a very likely process, because of the great reactivity and low standing concentration of the intermediates. Mechanisms that involve an attack of the nitrene on its precursor, or on some other long-lived intermediate in the solution, seem much more probable. It is significant that in carbene chemistry there are very few reaction products that can be ascribed to dimerization of free carbenes in solution.

Azo compounds can be formed in good yield from the thermolysis of some aryl azides, or by the reaction of nitro compounds with transition-metal oxalates:

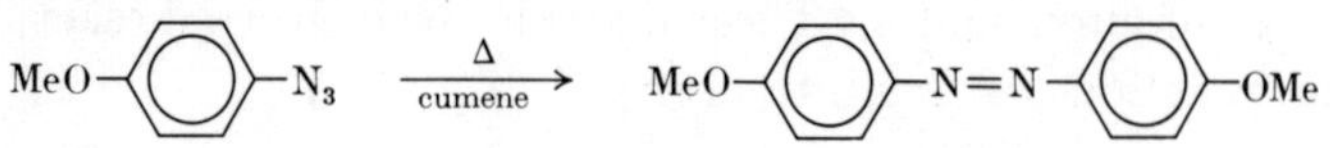

A nitrene–metal complex has been suggested in the latter reaction.[14] With the azides, an attack of the nitrene on the azide followed by loss of nitrogen, could account for the product, though again there is no direct evidence to support such a mechanism:

$$R\overset{+}{N}-N=\overset{-}{N} \quad :\ddot{N}R \longrightarrow RN=N-N=NR \longrightarrow RN=NR$$

Vapour-phase pyrolyses of azides give very good yields of the azo compounds if no intramolecular C–H insertions are possible.[15]

Tetrazenes, the formal dimers of aminonitrenes, are often the major products in the oxidations of 1,1-disubstituted hydrazines, especially at high concentrations of the hydrazine:

$$\underset{Ph}{\overset{Me}{>}}N-NH_2 \xrightarrow{Pb(OAc)_4} \underset{Ph}{\overset{Me}{>}}N-N=N-N\underset{Me}{\overset{Ph}{<}}$$

$$\underset{}{\bigcirc}N-NH_2 \xrightarrow{HgO} \underset{}{\bigcirc}N-N=N-N\underset{}{\bigcirc}$$

Other mechanisms are again more likely than simple dimerization of the free or complexed nitrenes; for example, the nitrene could attack the hydrazine to form a tetrazane, which would be easily oxidized to the tetrazene in the reaction conditions:

$$R_2N-\ddot{N}: + H_2N-NR_2 \longrightarrow R_2N-NH-NH-NR_2 \xrightarrow{[O]} R_2N-N=N-NR_2$$

7–6 Other reactions of aminonitrenes

Oxidations of 1,1-disubstituted hydrazines, which probably involve aminonitrenes as intermediates, have been intensively studied in the last few years.[13,16] Although tetrazene formation is the 'normal' reaction, it is far from being the only one. Rearrangements of the nitrenes and their addition to olefins have already been noted. Another important reaction pathway was reported by Busch and Weiss in 1900: 1,1-dibenzylhydrazine was oxidized by mercuric oxide to dibenzyl:

$$\underset{PhCH_2}{\overset{PhCH_2}{>}}N-NH_2 \xrightarrow{HgO} PhCH_2CH_2Ph$$

This reaction has since been extended to many other hydrazines where the 1,1-substituents are capable of existing as fairly stable radicals. The evidence shows that the products are not derived from the tetrazene, nor from an azo compound $PhCH_2N=NCH_2Ph$. The reaction is intramolecular: with unsymmetrical hydrazines no crossed products are found. In cyclic systems the reaction is virtually stereospecific:

This suggests that the loss of nitrogen from the nitrene and ring closure is concerted; however, an optical isomer of $\alpha\alpha'$-dimethyl-1,1-dibenzylhydrazine gave 2,3-diphenylbutane that was only partly optically active. A 'solvent-cage' fragmentation is a possible mechanism.

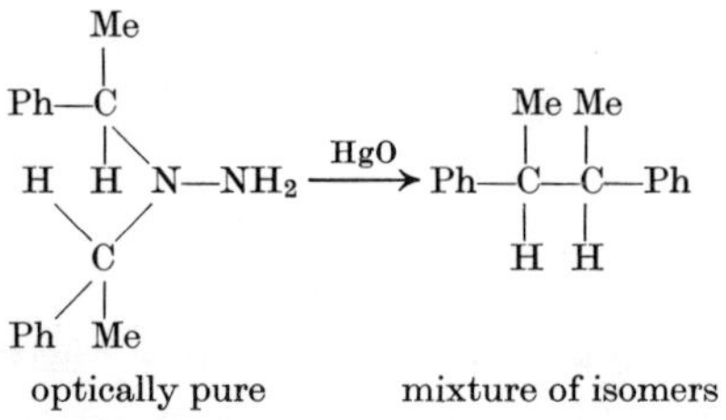

optically pure   mixture of isomers

Other hydrocarbon products are sometimes obtained in these reactions, which indicate that there is probably a diradical intermediate involved. Thus, the nitrene [7] derived from N-amino-dihydroisoindole gives the same hydrocarbon mixture as does 2-thia-indane dioxide [8] on thermolysis: a common diradical intermediate is likely:[17]

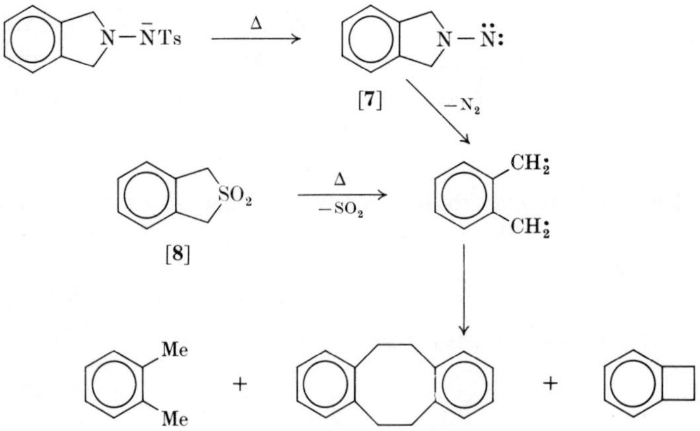

Deaminations are also fairly common in oxidations of 1,1-disub-stituted hydrazines (N-aminopyrrole and N-aminocarbazole are converted into pyrrole and carbazole, for example), but it is not certain whether these reactions involve aminonitrenes. The factors that decide the precise course which these interesting hydrazine oxidations are likely to follow are still not properly understood.

7–7 Reactions of nitrenes with nucleophiles

As electrophilic intermediates, nitrenes are expected to react readily with nucleophiles, and several of the reactions outlined already, such as those with amines, alcohols, and azides, are examples of nucleophilic attack on nitrenes. Other nucleophiles that can be used as nitrene traps include tertiary phosphines and phosphites, and sulphides. Thus, nitrenes, generated by the reaction of aromatic nitro compounds with triethyl phosphite, may be trapped by excess of the phosphite:[18]

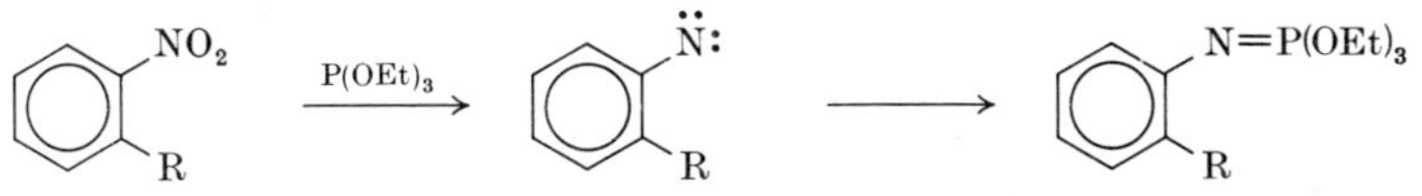

Similarly, nitrenes can be trapped by dimethyl sulphide or by dimethyl sulphoxide in solution:[19]

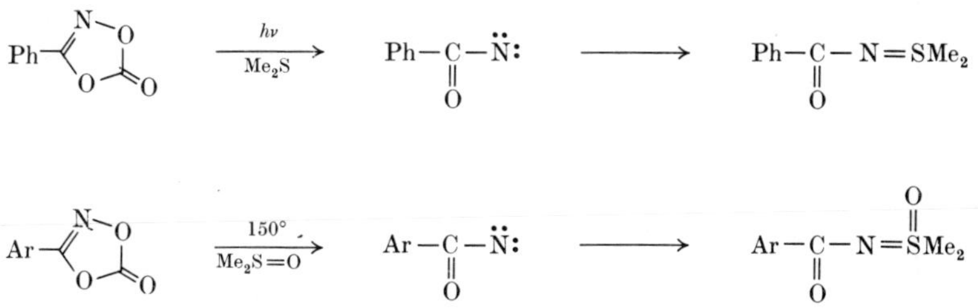

The latter is particularly useful as it is often a very suitable solvent in which to generate the nitrene.

The formation of phosphineimines and sulphoximines from nitrenes and phosphines or dimethyl sulphoxide is reversible; these compounds may regenerate nitrenes on photolysis.

7–8 Transition-metal complexes

Nitrene complexes of transition metals are suggested as intermediates in several reactions: the decomposition of sulphonyl azides catalysed by copper[5] and the reduction of nitro compounds with transition-metal oxalates[14] are examples.

Di-iron noñacarbonyl, $Fe_2(CO)_9$, catalyses the decomposition of azides. With phenyl azide, an iron–nitrene complex [9] is among the products.[20] $Fe_3(CO)_{12}$ also catalyses the decomposition of azides, and gives azobenzene from phenyl azide. Methyl azide with $Fe_2(CO)_9$ gives, among other products, a complex believed to have structure

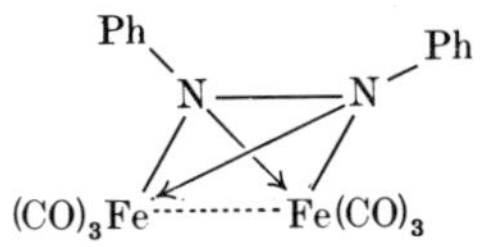

[9] 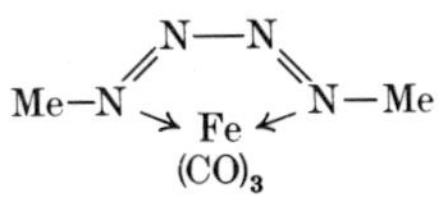   [10]

[**10**]. This complex [**10**] of the unknown tetrazadiene system illustrates the remarkable stabilizing effect that transition metals can have, and also indicates the sort of intermediates that may be involved in nitrene reactions.

References

1. W. Lwowski, *Angew. Chem. Internat. Edn.*, 1967, **6**, 897.
2. A. G. Anastassiou, *J. Amer. Chem. Soc.*, 1968, **90**, 1527.
3. A. Mishra, S. N. Rice, and W. Lwowski, *J. Org. Chem.*, 1968, **33**, 481.
4. R. S. Atkinson and C. W. Rees, *Chem. Comm.*, 1967, 1230.
5. H. Kwart and A. A. Khan, *J. Amer. Chem. Soc.*, 1967, **89**, 1950.
6. W. Nagata, S. Hirai, K. Kawata, and T. Aoki, *J. Amer. Chem. Soc.*, 1967, **89**, 5045.
7. W. von E. Doering and R. A. Odum, *Tetrahedron*, 1966, **22**, 81.
8. G. R. Harvey and K. W. Ratts, *J. Org. Chem.*, 1966, **31**, 3907; A. Hassner and F. W. Fowler, *Tetrahedron Lett.*, 1967, 1545.
9. A. G. Anastassiou, *J. Amer. Chem. Soc.*, 1966, **88**, 2322.
10. G. Smolinsky and B. I. Feuer, *J. Amer. Chem. Soc.*, 1964, **86**, 3085.
11. R. A. Abramovitch, J. Roy, and V. Uma, *Canad. J. Chem.*, 1965, **43**, 3407.
12. P. A. S. Smith, in *Molecular Rearrangements*, Vol. 1, Chapter 8, ed. P. de Mayo, 1963, Interscience, New York.
13. C. G. Overberger, J. P. Anselme, and J. G. Lombardino, *Organic Compounds with Nitrogen–Nitrogen Bonds*, 1966, Ronald Press, New York.
14. R. A. Abramovitch and B. A. Davis, *J. Chem. Soc. (C)*, 1968, 119.
15. G. Smolinsky, *J. Org. Chem.*, 1961, **26**, 4108.
16. C. G. Overberger and S. Altscher, *J. Org. Chem.*, 1966, **31**, 1728 and references therein.
17. W. Baker, J. F. W. McOmie, and D. R. Preston, *J. Chem. Soc.*, 1961, 2971.
18. R. J. Sundberg, *J. Amer. Chem. Soc.*, 1966, **88**, 3781.
19. J. Sauer and K. K. Mayer, *Tetrahedron Lett.*, 1968, 319.
20. M. Dekker and G. R. Knox, *Chem. Comm.*, 1967, 1243. See also R. J. Doedens, *ibid*, 1968, 1271.

Problems

7–1   Suggest mechanisms for the following reactions of azides:

(a)

(J. H. Boyer and D. Straw, *J. Amer. Chem. Soc.*, 1953, **75**, 2683).

(b)

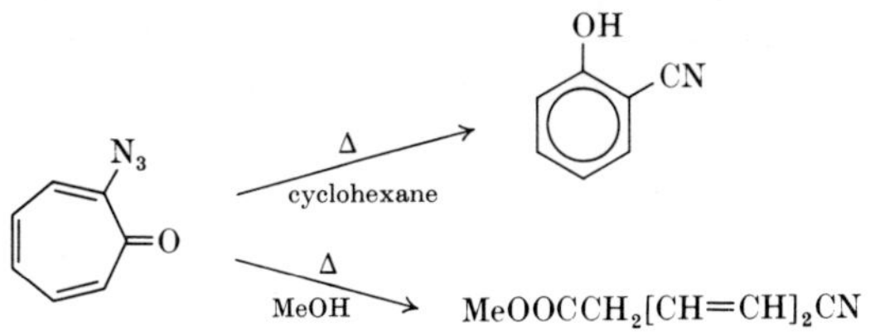

(J. D. Hobson and J. R. Malpass, *Chem. Comm.*, 1966, 141).

7–2 The following deoxygenations are accompanied by rearrangement. Suggest possible mechanisms for them.

(a)

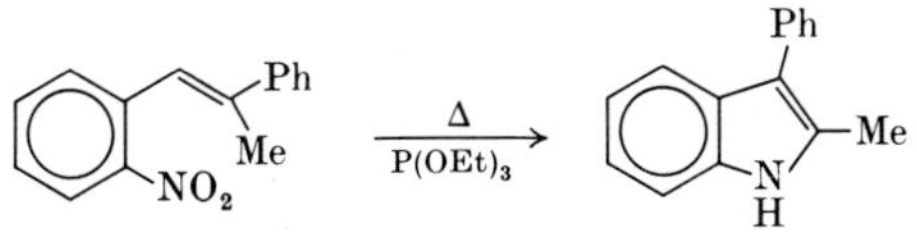

(R. J. Sundberg and T. Yamazaki, *J. Org. Chem.*, 1967, **32**, 290).

(b)

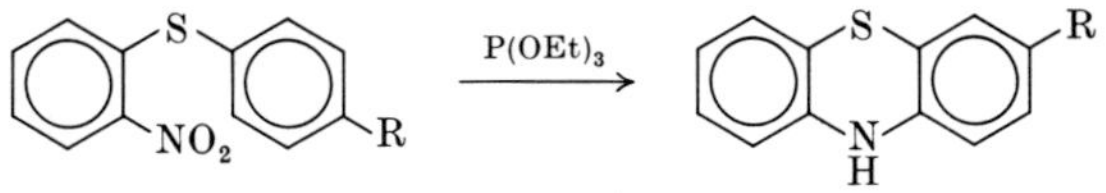

$$(R = Me, Cl)$$

(J. I. G. Cadogan, S. Kulik, and M. J. Todd, *Chem. Comm.*, 1968, **736**).

(c)

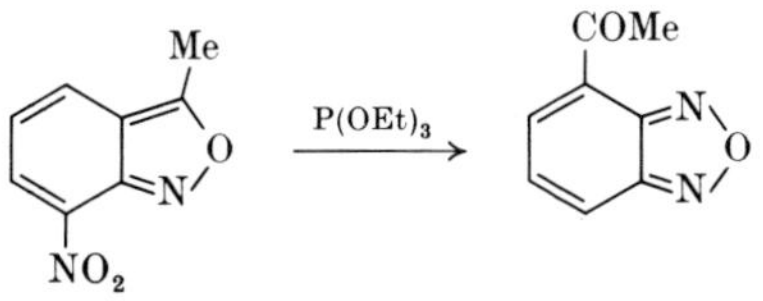

Suggest two mechanisms and an experiment to distinguish them.

(A. J. Boulton, I. J. Fletcher, and A. R. Katritzky, *Chem. Comm.*, 1968, 62).

(d)

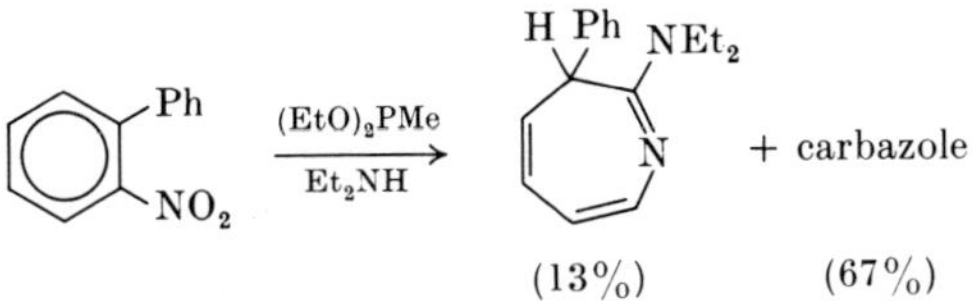

$$(13\%) \qquad (67\%)$$

(J. I. G. Cadogan, *Quart. Rev.*, 1968, **22**, 231).

7–3   Nitrenes react with dimethyl sulphoxide to give 1:1 adducts (Section 7–7). In contrast, carbenes react with dimethyl sulphoxide to give carbonyl compounds by abstraction of oxygen (see Section 6-3 and R. Oda, M. Mieno, and Y. Hayashi, *Tetrahedron Lett.*, 1967, 2363). Suggest a possible explanation for this difference in behaviour.

7–4   In the reaction:

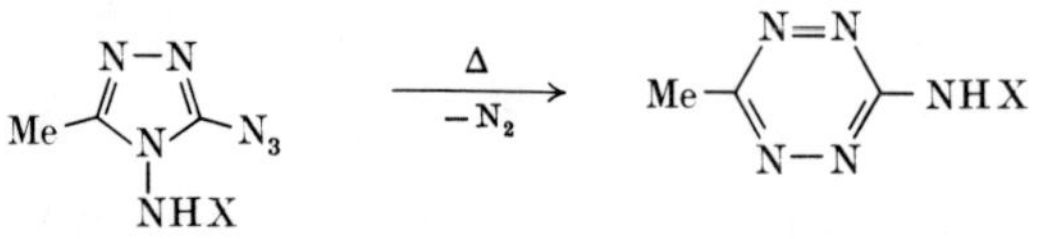

the azides (X = H and X = COMe) decompose at approximately the same rate. Suggest a mechanism.

(H. H. Takimoto and G. C. Denault, *Tetrahedron Lett.*, 1966, 5369).

Reactions
of arynes

8–1 Introduction

Arynes characteristically show properties expected of highly reactive acetylenes; they undergo cycloaddition reactions with conjugated dienes and with 1,3-dipolar systems very readily. In other reactions they are more akin to singlet carbenes and nitrenes; they react with nucleophiles of all kinds. Although they could in principle act as nucleophiles and react with electrophilic systems, examples of this type of reaction are rare. Radical reactions are also practically unknown.

8–2 Cycloadditions

Diels–Alder reactions. One of the most important and useful reactions of benzyne is the Diels–Alder reaction with conjugated dienes:

Benzynes give adducts with virtually all dienes that are capable of undergoing the Diels–Alder reaction, and will even add to systems not normally regarded as dienes at all, such as thiophen and benzene. Reactions of this type are often used as evidence for the intermediacy of benzyne. Since the reaction was first discovered by Wittig in the 1950s, a few dienes have come to be regarded as typical 'benzyne traps', mainly because they react very readily and can give high yields of adducts that are easily identified. Some common benzyne traps, and the products they give, are shown in Table 1. Some of the products, such as those of tetracyclone and of 2-pyrone, are not the primary Diels–Alder adducts but result from the loss of a bridging group, such as CO or CO_2, from the adduct, with the formation of a fully aromatic system.

Some of these adducts are important compounds in their own right, and this is usually the best way to make them. The reaction has been extended to a large number of other dienes, and to other arynes. Further synthetic applications are given in Chapter 9. The source of the aryne is usually not important unless a strong base, which is also

Table 1. Some common benzyne traps, and the products they give

Diene	*Benzyne adduct*	*Remarks*
 Furan	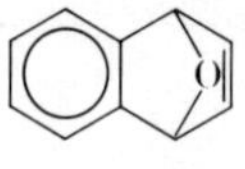 1,4-Dihydronaphthalene endoxide	Low-boiling furan is easy to use and to remove in solution reactions; the adduct is readily converted into 1-naphthol with acid
 Cyclopentadiene	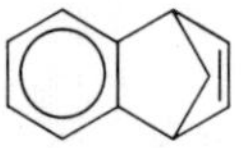 Benzonorbornadiene	The adduct is a thermally stable liquid that forms a crystalline adduct with phenyl azide
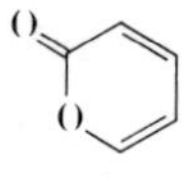 2-Pyrone	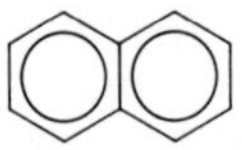	The primary Diels–Alder adduct decarboxylates spontaneously
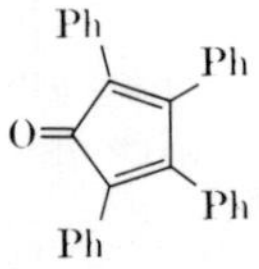 Tetraphenylcyclopentadienone ('tetracyclone')	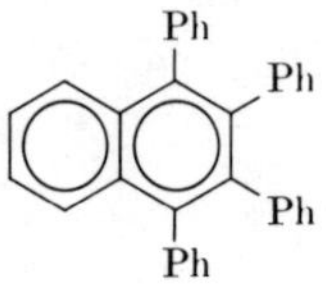	A very useful trap for following reactions, because the tetracyclone is highly coloured and the adduct colourless
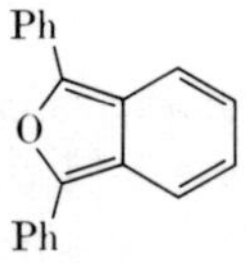 1,3-Diphenylisobenzofuran	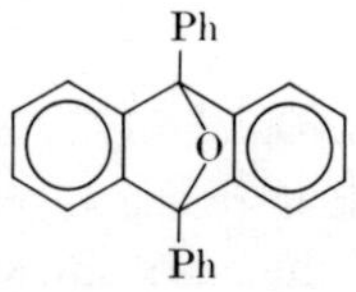 9,10-Dihydro-9,10-diphenyl anthracene endoxide	
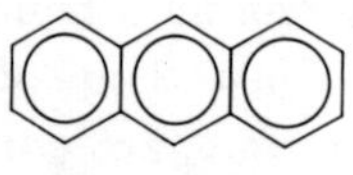 Anthracene	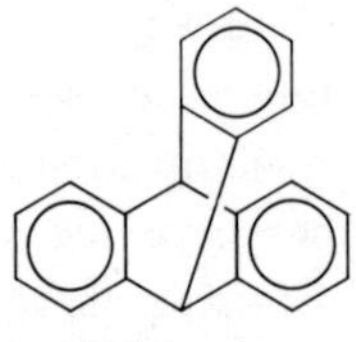 Triptycene	Benzyne addition across the terminal ring may also be observed with substituted anthracenes[1]

a strong nucleophile, is used to generate the aryne; in such reactions, nucleophilic addition is usually faster than the cycloaddition (Section 8–3).

The scope of the Diels–Alder reaction is wider with arynes than with stable olefins and acetylenes. Thus, pyrrole derivatives, which do not normally act as dienes in the Diels–Alder reaction, can add to arynes, though the yields are low.[2] Tetrafluorobenzyne is better than benzyne itself in these reactions; it has also been added to thiophen in the first observed Diels–Alder reaction of thiophen.[3]

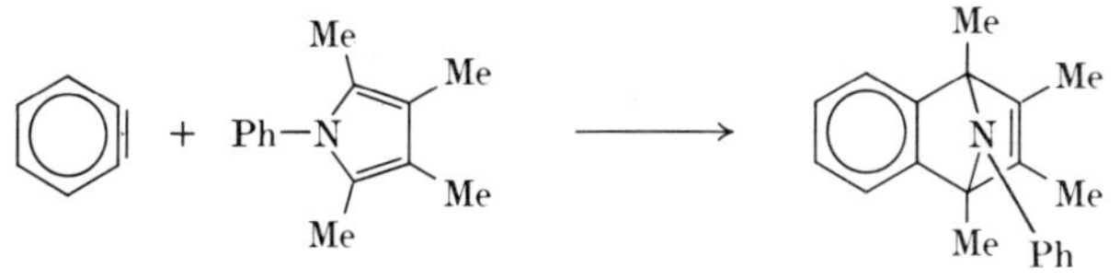

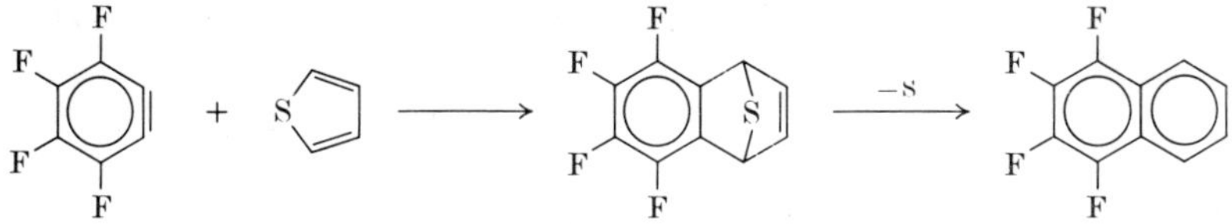

Benzyne and halogenated arynes can also add to benzene and its derivatives to give the Diels–Alder adducts, often in surprisingly good yields. With benzyne and benzene, other products are formed besides the 1,4-adduct. Some examples of the reaction are shown:

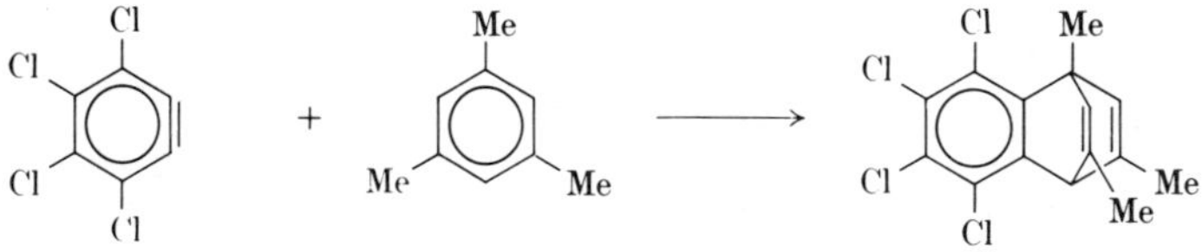

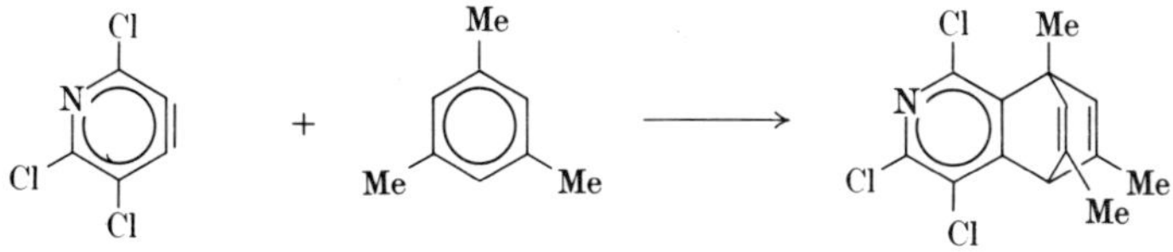

Again, these reactions are synthetically important because they give bridged ring systems that are otherwise inaccessible. We have noted that, in some of these adducts, the bridging group is spontaneously lost to give an aromatic product, but in others it is retained. The driving force for the extrusion of the bridge is clearly the aromatization of the organic fragment; it is also linked with the thermodynamic stability of the group that is lost. Thus, bridging groups that are spontaneously lost are $-N{=}N-$, $-\overset{\parallel}{\underset{O}{C}}{-}O-$, $-\overset{\parallel}{\underset{O}{C}}-$; the groups

$-CH{=}CH-$ and $-S-$ can be lost by heating the appropriate adducts, and groups $-CH_2-$, $-NR-$, and $-O-$ are not normally extruded. A comprehensive survey of extrusion reactions is available.[7]

   1,2-*Cycloadditions*. There is no clear pattern in the reactions of arynes with simple olefins, and very little is known about reactions with acetylenes. However, several strained olefins have given cyclobutane derivatives formed by 1,2-cycloaddition of the benzyne to the double bond:

Only the *exo* adduct is formed in this reaction between norbornene and benzyne;[8] the addition is possibly concerted. Similar reactions occur with terminal acetylenes, but the benzocyclobutene derivatives are unstable and may dimerize or react with more benzyne. Thus, benzyne from benzenediazonium-2-carboxylate adds to phenyl-acetylene to give the dibenzocyclo-octatetraene derivative [1] in quite good yield:[9]

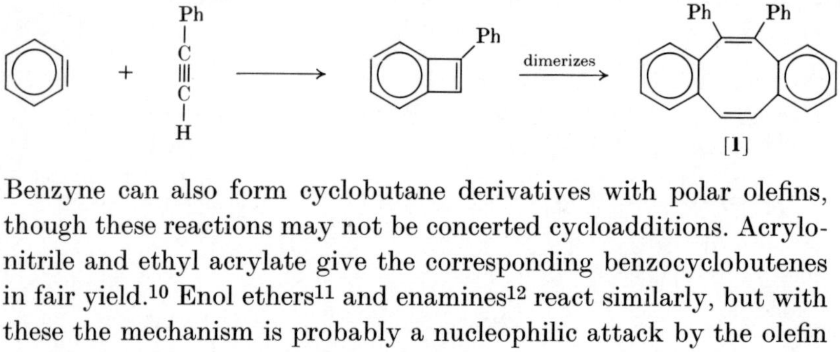

Benzyne can also form cyclobutane derivatives with polar olefins, though these reactions may not be concerted cycloadditions. Acrylonitrile and ethyl acrylate give the corresponding benzocyclobutenes in fair yield.[10] Enol ethers[11] and enamines[12] react similarly, but with these the mechanism is probably a nucleophilic attack by the olefin on the aryne followed by ring-closure:

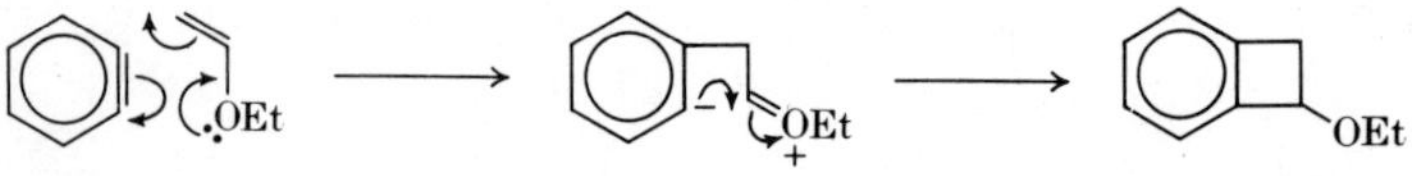

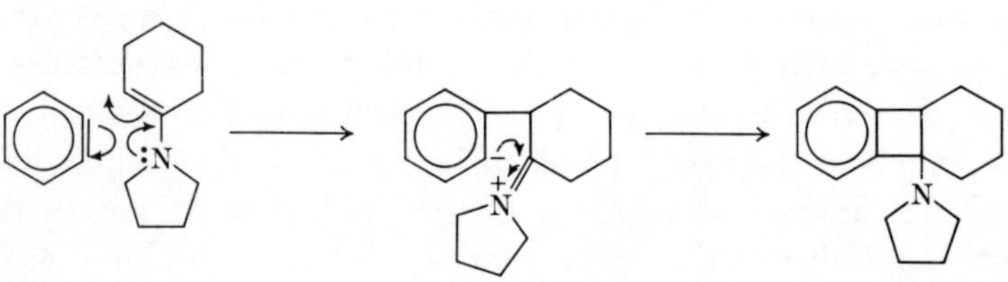

Ene reactions. Olefins with an allylic hydrogen can react with activated olefins and acetylenes by a concerted cycloaddition analogous to the Diels–Alder reaction. This is sometimes called the ene reaction. Several ene reactions have been observed with benzyne and allylic systems; some examples are shown:[8,13]

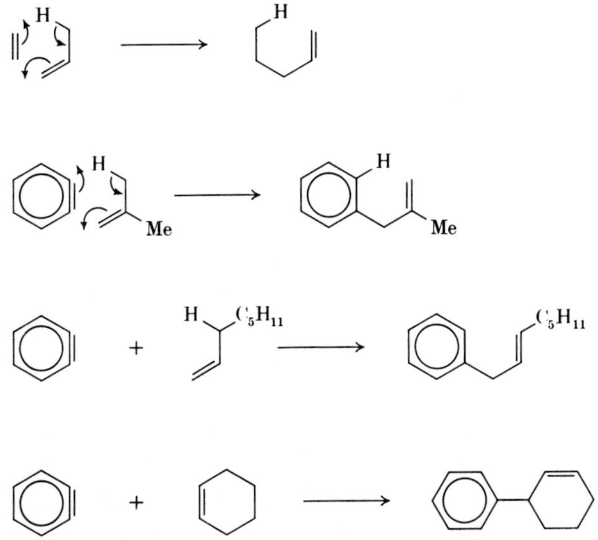

In these reactions there are usually other products as well, and the overall yields may not be very high.

1,3-*Dipolar additions.* Azides and diazo compounds can add as 1,3-dipoles to benzyne, just as they do to reactive olefins and acetylenes. The products are derivatives of benzotriazole[14] and of indazole:[15]

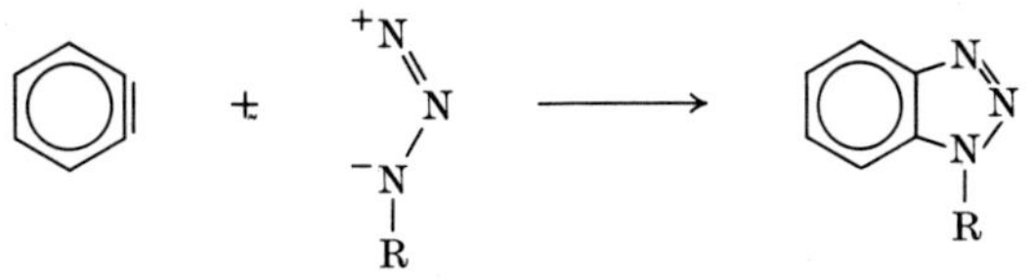

(R = alkyl, aryl, benzoyl, benzenesulphonyl)

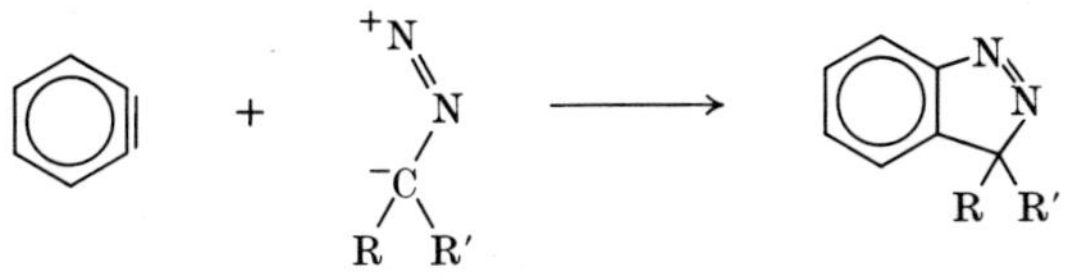

Similarly, benzonitrile oxide gives 3-phenylbenzisoxazole[16] [2]:

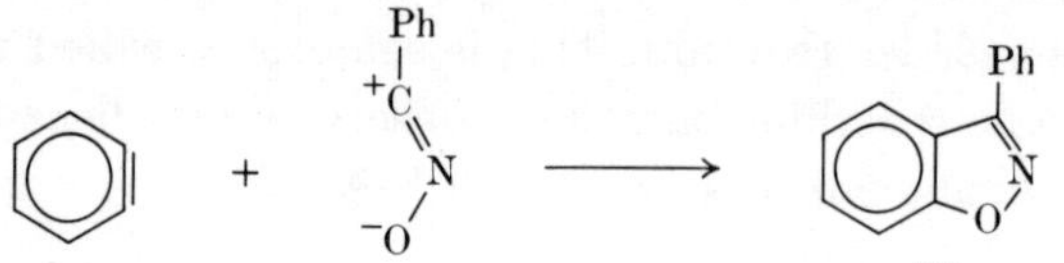

[2]

The reaction is a useful way of making these fused heterocyclic systems. The benzyne is best generated from benzenediazonium-2-carboxylate or a similar precursor.

8–3 Reactions with nucleophiles

Arynes react very readily with nucleophiles of all kinds. When the aryne is generated by the action of a base on a halogeno compound, some or all of the products are usually formed by the addition of the base to the aryne. For this reason, the Diels–Alder and related reactions characteristic of arynes in the presence of dienes cannot usually be carried out if the aryne is generated by potassamide in liquid ammonia or by similar routes. On the other hand, the 1,4-cycloaddition can compete successfully with the addition of poorer nucleophiles such as carboxylic acids, or even water or methanol, if they are present only in low concentrations.

Water and alcohols, primary and secondary amines, and thiols add to benzyne to give corresponding phenylated compounds. The mechanism is probably a stepwise one: addition of the nucleophile followed by proton transfer:

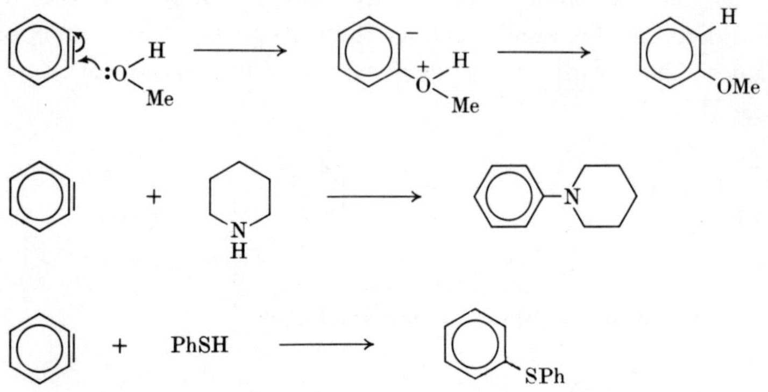

Phenyl esters can similarly be obtained from carboxylic acids, though not usually in very high yield. The anions of these nucleophiles react similarly.

Carbanions add to benzyne very readily. Methods of forming benzyne that involve phenyl anions as intermediates can lead to biphenyls and terphenyls as by-products:

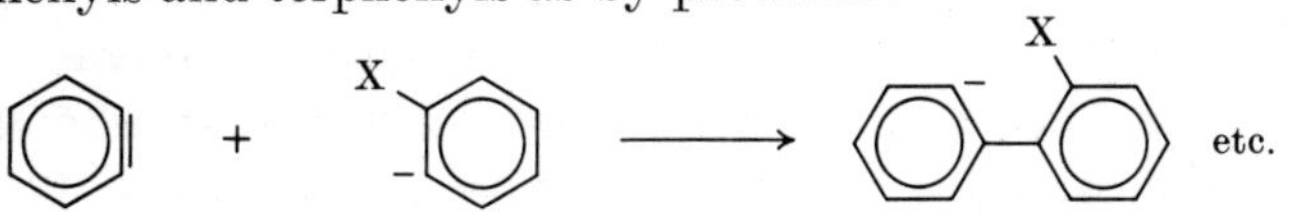

etc.

Addition of phenyl-lithium to arynes is a good way of synthesizing biphenyls. Similarly, some of the other carbanion additions to benzyne are synthetically useful; for example, the addition of malonate anions to benzyne provides a way of making phenylmalonic esters that are not easy to get by standard routes. The most common source of benzyne in these reactions is chlorobenzene with sodamide in liquid ammonia; the required carbanion is also readily generated in these conditions. In general, the more polarizable the nucleophile, the more readily it adds to benzyne.

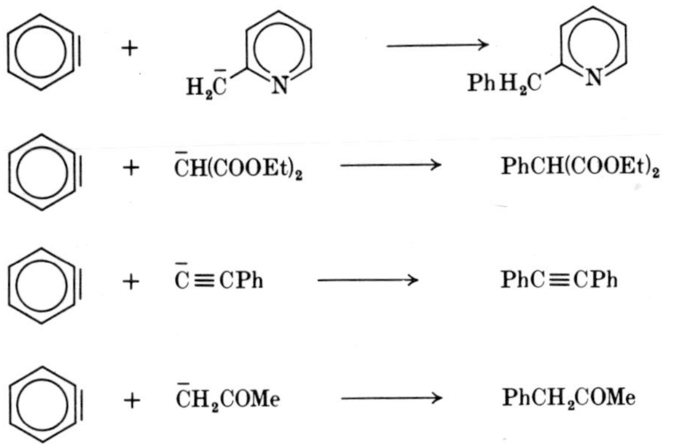

A useful synthetic development of the nucleophilic addition reaction has been the generation of arynes containing nucleophiles in an adjacent side chain. The intramolecular addition of the nucleophile to the aryne produces a new fused-ring system:

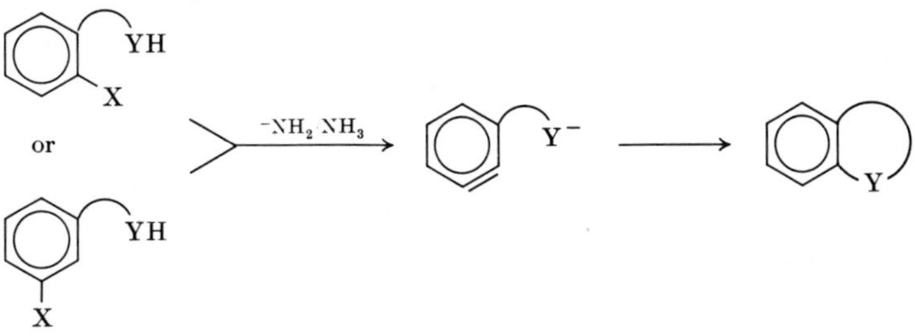

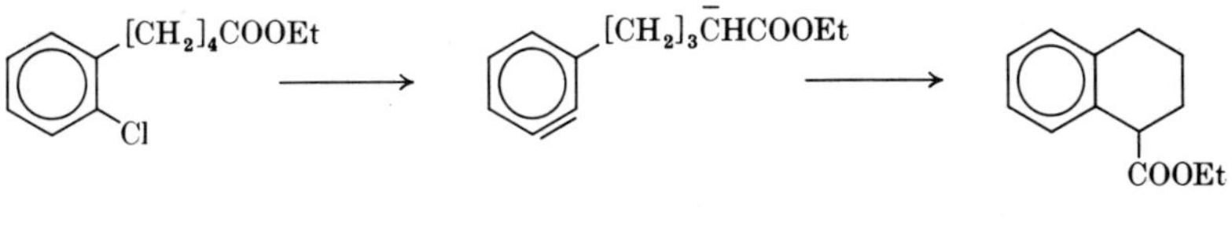

Reactions of this sort have been used to synthesize compounds that were otherwise difficultly accessible; more examples are given in Chapter 9.

Arynes can also react with sulphides and tertiary amines and phosphines, where the first intermediate must be a zwitterion. The zwitterionic intermediates can then react in various ways. If the nucleophile carries an alkyl group that has a β-hydrogen, then the usual reaction is an elimination of an olefin (a kind of intramolecular Hofmann elimination) through a cyclic transition state. Otherwise, the zwitterion can rearrange to an ylid, which may itself rearrange or react intermolecularly. With ethers, which are formally similar, the attack on benzyne is probably reversible, because the products expected from the initial betaine adduct are rarely found. 1,2-Dimethoxyethane does give a betaine, however, which can be trapped. Examples of these reactions are shown:

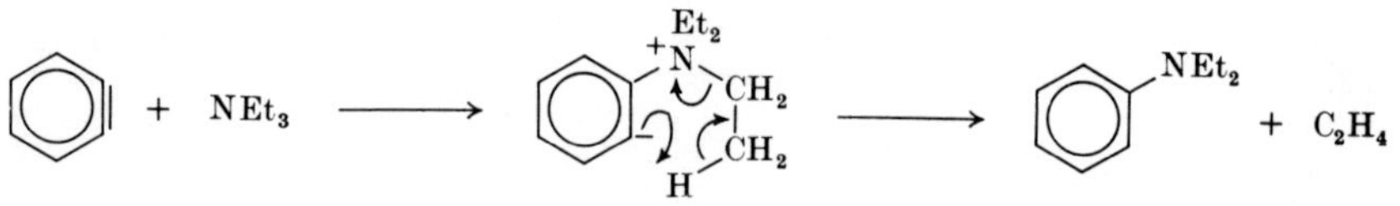

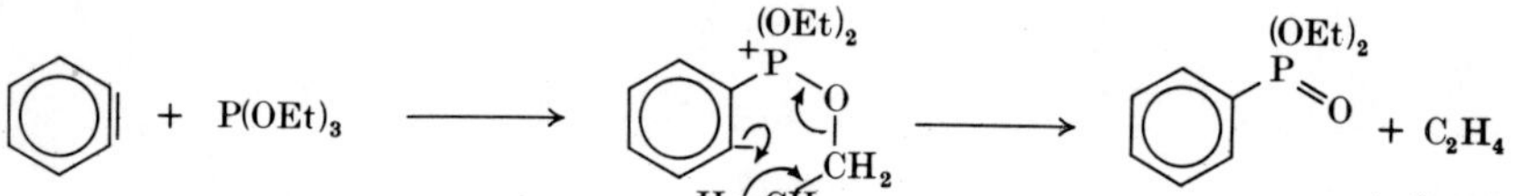

(ref. 17)

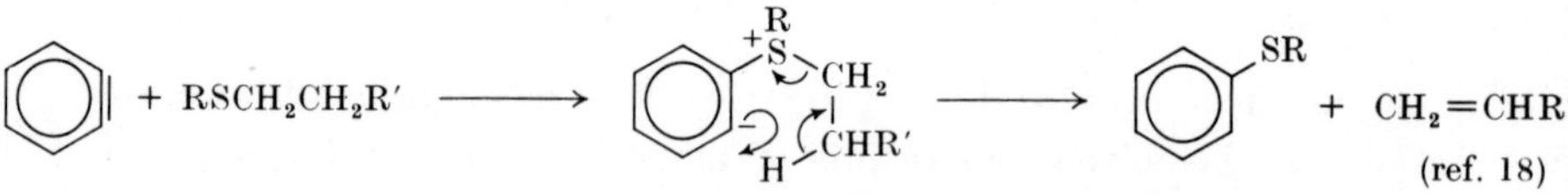

(ref. 18)

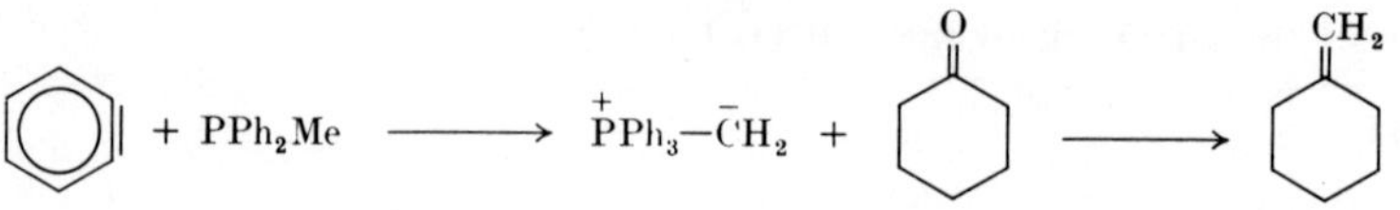

(ref. 19)

(Wittig reaction)

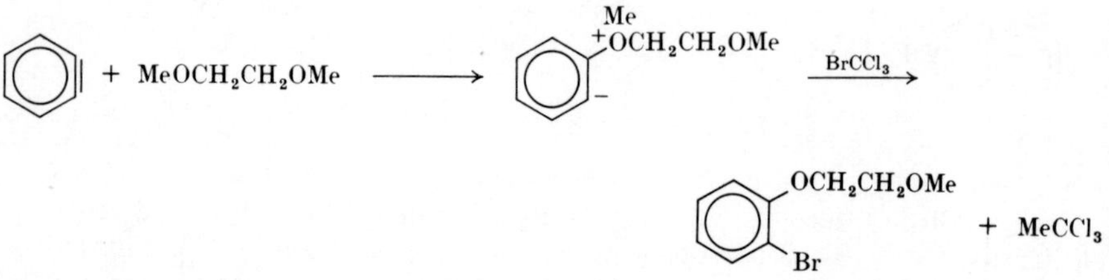

Nucleophilic attack on unsymmetrical arynes. With unsymmetrical arynes there are two possible products of nucleophilic addition, which result from an attack of the nucleophile at either end of the aryne bond:

The ratio of the isomers formed is dependent mainly on the substituent, R, which will preferentially stabilize one of the two possible transition states in the nucleophilic addition. In the attack of a nucleophile X^- on a 3-substituted benzyne, for example, the two possible transition states can be represented as follows:

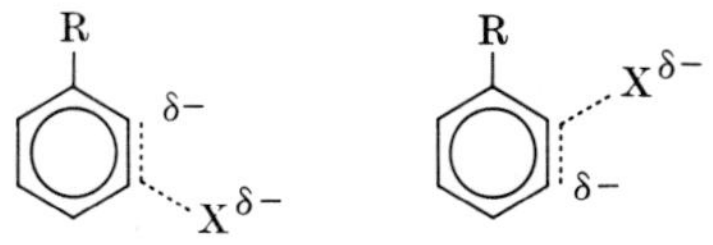

These will be stabilized or destabilized inductively by the substituent R. If R is inductively electron-withdrawing, it will preferentially stabilize the developing negative charge at the *ortho* position and so favour *meta* attack by the nucleophile. Assuming that the non-bonding orbitals of the aryne are orthogonal to the π-electron system, the conjugative effect of the substituent can only come into play once the new bond to the nucleophile is formed, so it is likely to be much less important than the inductive effect. A conjugatively electron-withdrawing group (CN, NO_2) will stabilize the new substituent at the *ortho* position, whereas a conjugatively releasing group (alkyl, OR, NR_2) will stabilize it at the *meta* position. These substituent effects can be illustrated for a 3-substituted benzyne by predicting the positions of attack and comparing the predictions with the experimental results:

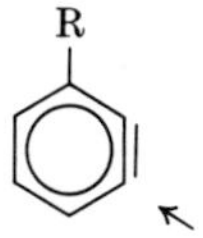

R inductively withdrawing $(-I)$ R inductively releasing $(+I)$
or R conjugatively releasing $(+M)$ *or* R conjugatively withdrawing $(-M)$

The results in Table 2 suggest that the inductive effect of the substituent is the factor that determines the preferred position of attack, with the conjugative effect perhaps slightly influencing the product

Table 2. Amination (KNH₂/NH₃) of 3-substituted benzynes[20]

Substituent R		Meta:ortho product ratio
CN	$(-I, -M)$	$85-90:10-15$
F	$(-I, +M)$	$99-100:0-1$
OMe	$(-I, +M)$	$95-100:0-5$
Me	$(+I, +M)$	$45:55$
NMe₂	$(-I, +M)$	$95-100:0-5$
O⁻	$(+I, +M)$	$10-15:85-90$
NH⁻	$(+I, +M)$	$5-10:90-95$

ratio. With 4-substituted arynes the substituent effect is generally smaller, presumably because the inductive influence of the substituent is weaker.

The ratio of isomers may also depend on the nature of the attacking nucleophile. The better the nucleophile, or the higher its concentration, the less selective the addition is likely to be. This is illustrated by the isomer ratios for the addition of methanol to 3,4-dehydrochlorobenzene:[21]

	Ratio	m- : p-chloroanisole
	2M-NaOMe/MeOH	$1:2$
	0·1M-NaOMe/MeOH	$1:5$
	MeOH	$1:6$

Methoxide ion is a better nucleophile than methanol, so the selectivity of the addition increases as its concentration decreases.

Steric effects may be important in determining isomer ratios if the nucleophile is a very bulky one. For example, the 1-position of 1,2-naphthalyne is hindered to attack by bulky amines such as di-isopropylamine because of interaction with the *peri*-hydrogen on the adjacent ring. The hindrance is even greater with 9,10-phenanthryne:[22]

8–4 Reactions with electrophiles and radicals

The introduction of benzyne precursors like benzenediazonium-2-carboxylate, benzothiadiazole 1,1-dioxide, and 1-aminobenzotriazole has made the reactions of benzyne with electrophiles feasible; they were naturally excluded from the early work when benzyne could only be generated in strongly basic conditions. Reactions are now known in which benzyne, from the diazonium carboxylate, is attacked by typical electrophiles such as the halogens. With iodine, this is a useful synthesis of o-di-iodobenzene:[23]

Although this appears to be a typical electrophilic addition by the halogen, the reaction should be interpreted with caution. If benzyne is a stronger electrophile than the halogen, then the same product could be obtained from an electrophilic attack *on* the halogen by benzyne. However, electrophilic attack by the halogen seems more likely. With triethylboron benzyne reacts to give, after hydrolysis, phenylboronic acid, and here an electrophilic attack on benzyne is the only reasonable mechanism:

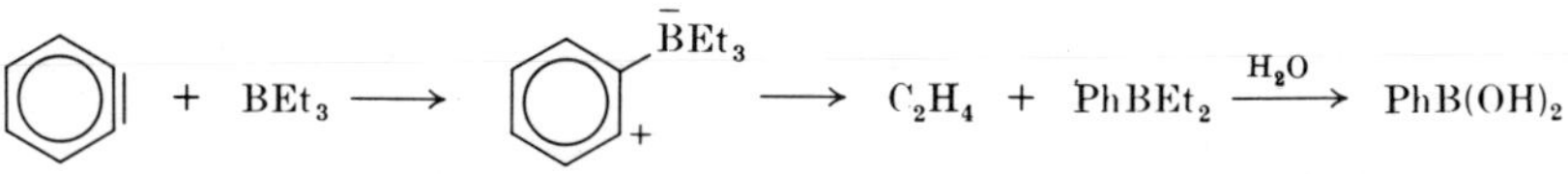

Very little is known about radical reactions of benzyne. Benzene has been found as a product of the photolysis of the benzyne precursor o-iodophenylmercuric iodide in cyclohexane and similar solvents; it may have been formed by hydrogen abstraction by triplet benzyne. Otherwise, evidence for radical character in benzyne is conspicuous by its absence.

8–5   Dimerization and trimerization

Biphenylene [**3**] and triphenylene [**4**], both stable hydrocarbons, are respectively the dimer and trimer of benzyne. Both are formed in

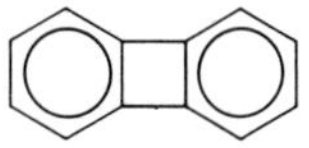

[**3**]

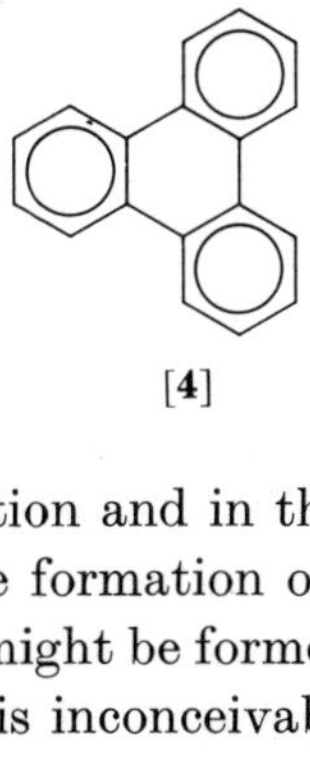

[**4**]

benzyne reactions, in solution and in the gas phase. Biphenylene is not an intermediate in the formation of triphenylene. Though it is possible that biphenylene might be formed by concerted dimerization of benzyne in solution, it is inconceivable that more than traces of triphenylene could be formed from a concerted trimerization. All the precursors of triphenylene in solution are organometallic intermediates. In such systems triphenylene can be formed by a stepwise reaction between benzyne and the metallated intermediates, and very high yields (up to 85%) of the hydrocarbon can be obtained. Biphenylene is sometimes a by-product. The following mechanism accounts for the formation of biphenylene and triphenylene:

(M = Li, MgBr)

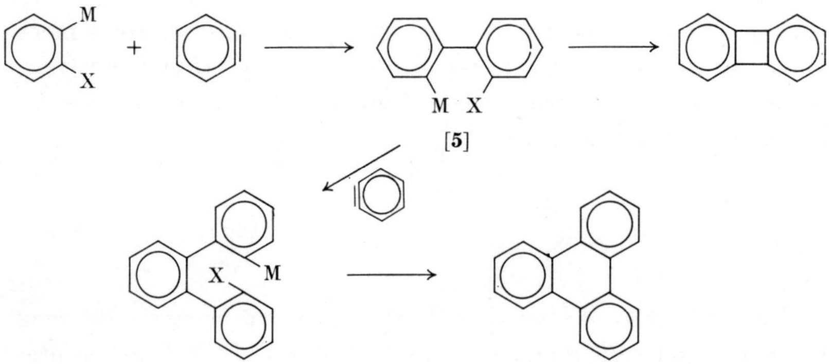

[5]

The intermediacy of the 2-metallated 2′-halogenobiphenyl [5] in the reaction can be shown by adding a substituted biphenyl derivative of the same sort: the unsymmetrically substituted triphenylene is among the products:

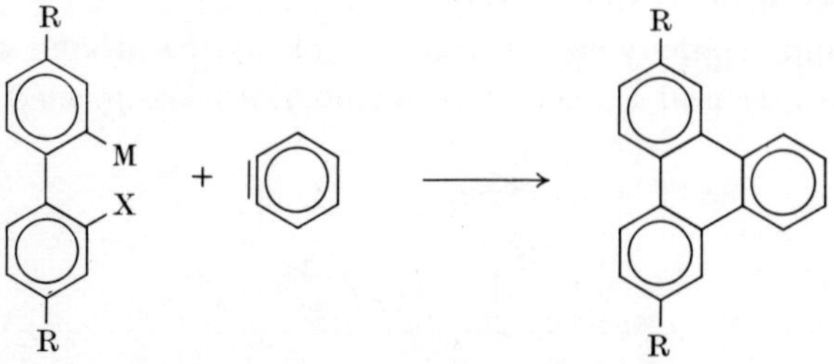

Triphenylene is formed to some extent in nearly every benzyne reaction involving these organometallic precursors, even in the presence of nucleophiles and diene traps. Only a little is ever formed in solution reactions of other benzyne precursors such as benzenediazonium-2-carboxylate, benzothiadiazole 1,1-dioxide, or 1-aminobenzotriazole. In the absence of a trap these give varying amounts of biphenylene. With lead tetra-acetate, 1-aminobenzotri-azole gives as much as 85% of biphenylene, and the dimerization can compete effectively with attack of weak nucleophiles on the benzyne. The reason is probably that the benzyne is generated rapidly and in high local concentration. The only other useful source of biphenylene in solution is benzenediazonium-2-carboxylate, which can give up to about 20% of the hydrocarbon when it is decomposed in hot dichloro-ethane. Again, presumably, benzyne is generated in high local concentration.[24] Benzothiadiazole 1,1-dioxide gives rather low yields of biphenylene in solution.

Gas-phase pyrolyses of the diazonium carboxylate, phthalic an-hydride, phthaloyl peroxide, benzothiadiazole dioxide, and several similar precursors give appreciable amounts of biphenylene, often with a little triphenylene. In the gas phase, small proportions of triphenylene may indeed come from a concerted trimerization of benzyne, or perhaps from an 2,2'-biphenyl diradical with benzyne. If the proportions of triphenylene are large, however, we can assume that it is formed through an organometallic intermediate.

Some of the more important benzyne precursors, and the propor-tions of biphenylene and triphenylene they give in different conditions, are shown in Table 3.

Substituted biphenylenes and triphenylenes can similarly be made from substituted benzyne precursors; for example:[25]

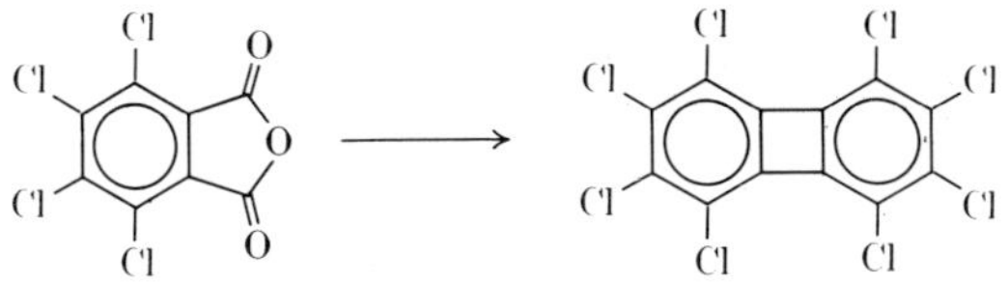

Unsymmetrical biphenylenes can be made by the co-oxidation of aminotriazole derivatives:

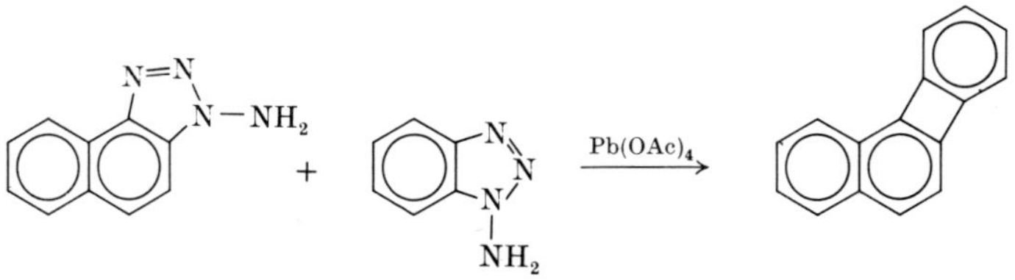

This is probably the best route to these hydrocarbons.[26]

Attempts to make dimers of 3,4-pyridyne and other hetarynes have so far failed; the heterocyclic analogues of biphenylene remain unknown.

Table 3. Some of the more important benzyne precursors, and the proportions of biphenylene and triphenylene they give in different conditions.

			% biphenylene	% triphenylene
		Ether/20° Flash pyrolysis	9 50	0 1
		Dichloroethane/80° Benzene/45° Flash photolysis	20 0–5 35	0 0 5
		With zinc/500°	20	10
		230° 600°	0 50	30 Some
		500–1100°	15	—
		600°	27	—
		Pb(OAc)$_4$/CH$_2$Cl$_2$	85	—

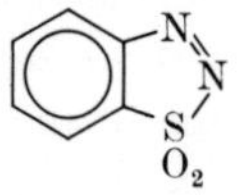

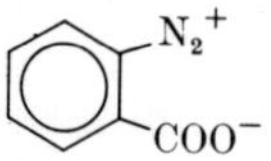

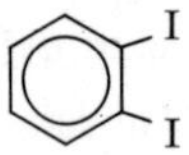

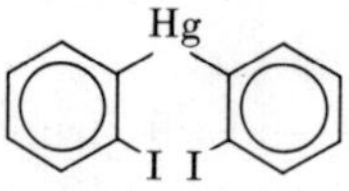

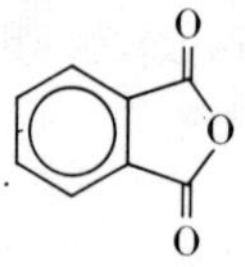

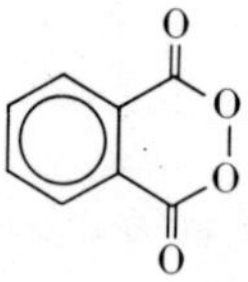

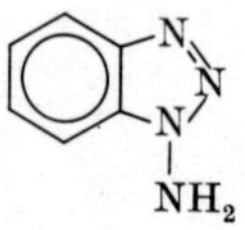

8–6   Transition-metal complexes

There have been several attempts to stabilize benzyne through a transition-metal complex. So far, attempts to do so by using derivatives of silver[27] and of platinum[28] have not succeeded, though Friedman noticed that even minute traces of silver ions had a dramatic effect on the product composition in the reaction of benzenediazonium-2-carboxylate with benzene; the yield of the 'normal' 1,4-adduct [6] is decreased at the expense of the products of formal insertion (biphenyl) and 1,2-cycloaddition (benzocyclo-octatetraene, [7]).

[6]
main product in the
absence of Ag$^+$

[7]
main products in the
presence of Ag$^+$

Attempts to synthesize the platinum complex [8] lead instead to complexes such as [9] in which the benzyne precursor is trapped before it can decompose to benzyne.[28,29] However, in vigorous conditions, the reactions of benzothiadiazole 1,1-dioxide are drastically altered by the presence of zerovalent platinum species. Triphenylene is formed in good yield, with no biphenylene (a reversal of the usual situation in solution decompositions of this precursor[29]).

An apparently successful attempt to obtain a benzyne–metal complex has been reported.[30] A black complex, assigned structure [10], is obtained from the reaction of tetracarbonylnickel with o-di-iodo-benzene (see Problem 8–4).

[8]

[9]

[10]

References

1. B. H. Klanderman, *J. Amer. Chem. Soc.*, 1965, **87**, 4649.
2. E. Wolthuis, W. Cady, R. Roon, and B. Weidenaar, *J. Org. Chem.*, 1966, **31**, 2009.
3. D. D. Callander, P. L. Coe, and J. C. Tatlow, *Chem. Comm.*, 1966, 143.
4. R. G. Miller and M. Stiles, *J. Amer. Chem. Soc.*, 1963, **85**, 1798.
5. H. Heaney and J. M. Jablonski, *Tetrahedron Lett.*, 1966, 4529.
6. J. D. Cook and B. J. Wakefield, *Tetrahedron Lett.*, 1967, 2535.
7. B. P. Stark and A. J. Duke, *Extrusion Reactions*, 1967, Pergamon Press, Oxford.
8. H. E. Simmons, *J. Amer. Chem. Soc.*, 1961, **83**, 1657.
9. M. Stiles, U. Burckhardt, and A. Haag, *J. Org. Chem.*, 1962, **27**, 4715.
10. T. Matsuda and T. Mitsuyasu, *Bull. Chem. Soc. Japan*, 1966, **39**, 1342.
11. H. H. Wasserman and J. Solodar, *J. Amer. Chem. Soc.*, 1965, **87**, 4002. For preliminary stereochemical investigations, see M. Jones and R. H. Levin, *Tetrahedron Lett.*, 1968, 5593; H. H. Wasserman, A. J. Solodar, and L. S. Keller, *ibid.*, 1968, 5597; L. Friedman, R. J. Osiewicz, and P. W. Rabideau, *ibid.*, 1968, 5735.

12. M. E. Kuehne. *J. Amer. Chem. Soc.*, 1962, **84**, 837.

13. G. Wittig and H. Dürr, *Annalen*, 1964, **672**, 55.

14. G. A. Reynolds, *J. Org. Chem.*, 1964, **29**, 3733.

15. W. Ried and M. Schön, *Chem. Ber.*, 1965, **98**, 3142; *Annalen*, 1965, **689**, 141.

16. F. Minisci and A. Quilico, *Chimica e Industria*, 1964, **46**, 428; *Chem. Abs.*, 1964, **60**, 15851h.

17. C. E. Griffin and N. T. Castellucci, *J. Org. Chem.*, 1961, **26**, 629.

18. V. Franzen, H.-I. Joschek, and C. Mertz, *Annalen*, 1962, **654**, 82.

19. D. Seyferth and J. M. Burlitch, *J. Org. Chem.*, 1963, **28**, 2463.

20. G. B. R. de Graaff, H. J. den Hertog, and W. C. Melger, *Tetrahedron Lett.*, 1965, 963.

21. J. F. Bunnett, *et al.*, *J. Amer. Chem. Soc.*, 1966, **88**, 5250.

22. T. Kauffmann, *et al.*, *Tetrahedron Lett.*, 1967, 2911, 2917.

23. L. Friedman and F. M. Logullo, *Angew. Chem. Internat. Edn.*, 1965, **4**, 239.

24. L. Friedman and A. Seitz, unpublished.

25. M. P. Cava, M. J. Mitchell, D. C. De Jongh, and R. Y. Van Fossen, *Tetrahedron Lett.*, 1966, 2947.

26. J. W. Barton and S. A. Jones, *J. Chem. Soc. (C)*, 1967, 1276.

27. L. Friedman, *J. Amer. Chem. Soc.*, 1967, **89**, 3071.

28. H. J. S. Winkler and G. Wittig, *J. Org. Chem.*, 1963, **28**, 1733; C. D. Cook and G. S. Jauhal, *J. Amer. Chem. Soc.*, 1968, **90**, 1464.

29. T. L. Gilchrist, F. J. Graveling, and C. W. Rees, *Chem. Comm.*, 1968, 821.

30. E. W. Gowling, S. F. A. Kettle, and C. M. Sharples, *Chem. Comm.*, 1968, 21.

Problems

8–1   The following transformations are all effected by benzyne. How can they be rationalized?

(a).

(E. Wolthuis, D. VanderJagt, S. Mels, and A. DeBoer, *J. Org. Chem.*, 1965, **30**, 190).

(b)

(J. C. Sheehan and G. D. Daves, *J. Org. Chem.*, 1965, **30**, 3247).

(c)

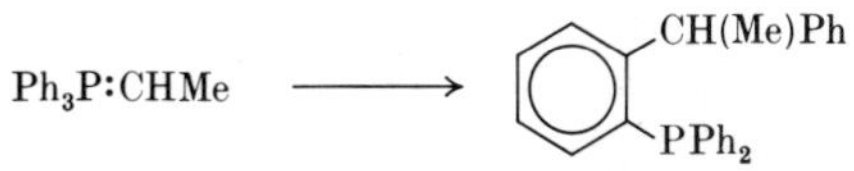

Ph$_3$P:CHMe ⟶

(E. Zbiral, *Tetrahedron Lett.*, 1964, 3963).

(d) PhNMe$_2$ ⟶ Ph$_2$NMe + Ph$_2$NEt

(A. R. Lepley, A. G. Giumanini, A. B. Giumanini, and W. A. Khan, *J. Org. Chem.*, 1966, **31**, 2051).

(e)

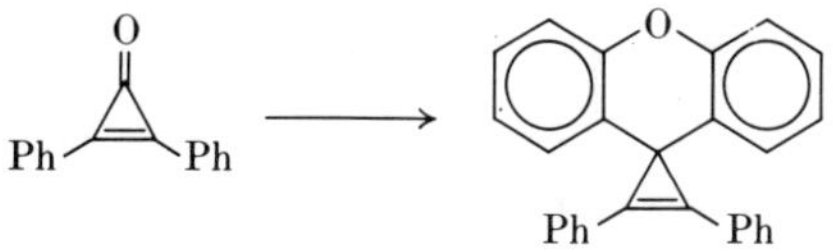

(J. Ciabattoni, quoted in *Dehydrobenzene and Cycloalkynes*, p. 205).

8–2   Benzyne, from benzenediazonium-2-carboxylate, adds non-stereo-specifically to propenyl *cis*- and *trans*-methyl ether, MeCH=CHOMe, to give the corresponding benzocyclobutenes. What, if anything, does this indicate about the mechanism of the cycloaddition and the structure of benzyne?

(I. Tabushi, R. Oda, and K. Okazaki, *Tetrahedron Lett.*, 1968, 3743; see also Ref. 11).

8–3   Suggest an explanation for the effect of silver ions observed in the reaction between benzyne (from the diazonium carboxylate) and benzene (Section 8–6). Compare your ingenuity with that of the discoverer! (Ref. 27.)

8–4   The structure of the nickel–benzyne complex [**10**] was assigned on the following evidence: elementary analysis for [NiC$_7$H$_4$I$_2$O]$_n$, a strong infra-red band at 1790 cm^{-1}, an A_2B_2 type of nuclear magnetic resonance spectrum ($J = 3 \cdot 5$ c/sec, $\delta = 14$ c/sec centred at τ 2·33) run in CH$_3$OD, and the mass spectrum. In the last, a parent peak was not observed, the heaviest ion being NiI$_2$C$_6$H$_4$$^+$; other ions produced included C$_6$H$_5$I$_2$$^+$ and ions derived from this by loss of H or I atoms. The compound was moder-ately stable in organic solvents giving brown solutions, which were instantly decolorized when shaken with water (ref. 30).
   Discuss this evidence critically.

9 Synthetic applications

9–1 Introduction

The development of the chemistry of reactive intermediates has had important implications for the practising organic chemist. It has introduced new synthetic routes to compounds that were previously difficult to make, or inaccessible. These in turn have influenced the theoretical development of the subject: many of the syntheses of non-benzenoid aromatic compounds, for example, are based on carbene reactions, and the study of the chemistry of these compounds has helped the development of the theory of aromaticity.

In this chapter we have selected just a few of the reactions described in the rest of this book, and have tried to illustrate some of their practical applications in organic synthesis.

9–2 Dihalogenocarbenes

The addition of dihalogenocarbenes to olefins provides a simple route to *gem*-dihalogenocyclopropanes, which are extremely useful synthetic intermediates.[1] In particular, they provide a means of extending a cyclic or acyclic carbon chain by one carbon atom. The reaction of the cyclopropanes with silver salts, first reported by Skell in 1958, is a convenient way of opening the cyclopropane ring:[2]

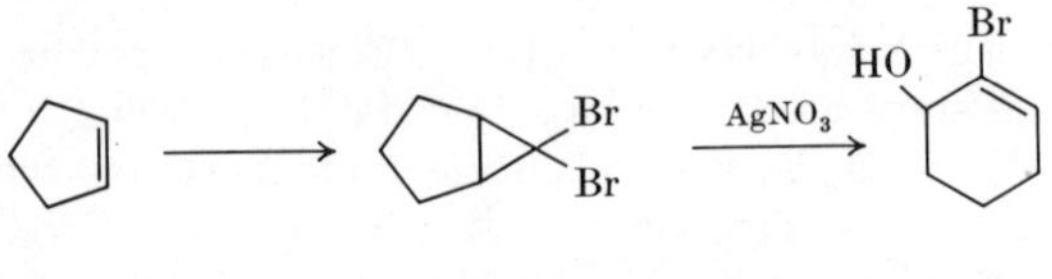

Similar reactions occur with acyclic olefins such as styrene and *cis*- or *trans*-butene. Some of these reactions have become useful syntheses of important ring systems. For example, tropone can be obtained in good yield from anisole by Birch reduction, addition of dibromocarbene to the more nucleophilic double bond, and treatment of the cyclopropane with aqueous silver nitrate:[3]

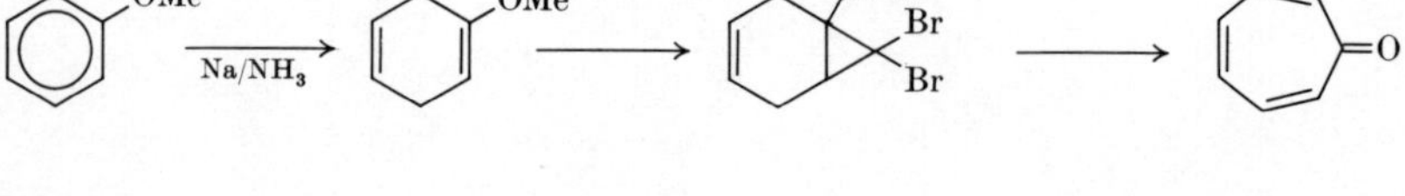

By a similar route, benzocyclobutenone derivatives can be obtained from *p*-dimethoxybenzene:[4]

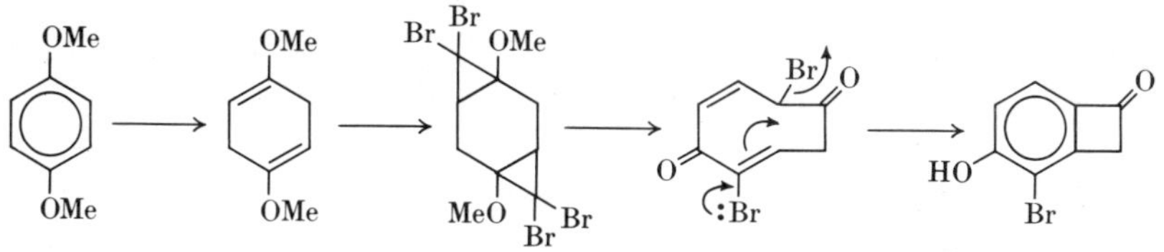

The ring-expansion can also be brought about by tertiary amines like pyridine and quinoline: Parham has applied this technique to the synthesis of a variety of cyclic systems. For example, dihydropyran can be converted into 3-chloro-5,6-dihydro-oxepine[5] [**1**], and 1-ethoxycyclododecene [**2**] is similarly ring-expanded in excellent yield:[6]

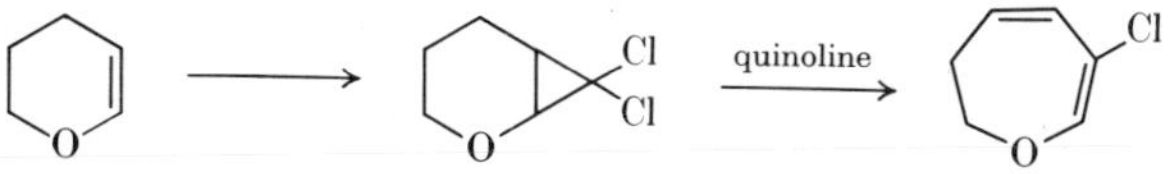

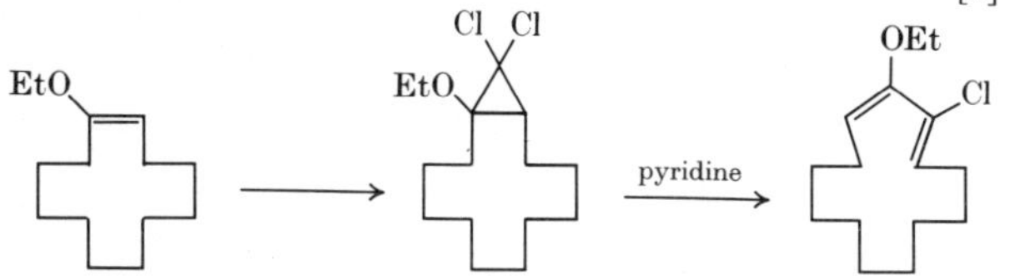

Similar reactions occur with aromatic carbocyclic and heterocyclic systems at bonds that are activated to electrophilic addition (Chapter 5). In these systems the dihalogenocarbene adducts are normally too unstable to be isolated, and the ring-expanded products are obtained directly. For example, the metacyclophane [**3**] is formed by the reaction of the corresponding indenes with dichlorocarbene:[7]

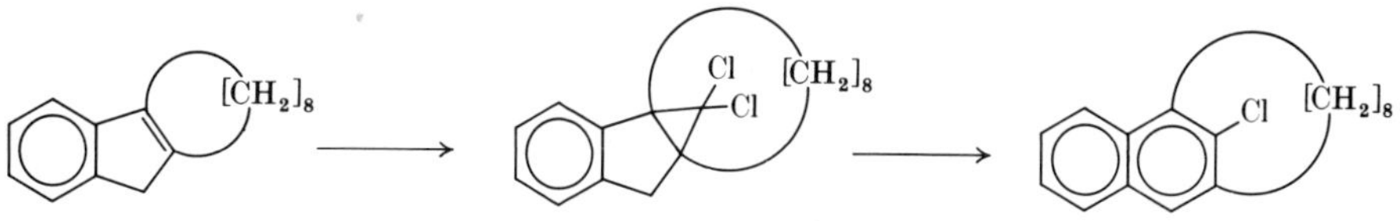

Another important use of *gem*-dihalogenocyclopropanes is in the synthesis of allenes. Olefins can be converted into allenes in good yield by adding dibromocarbene and treating the adduct with a lithium alkyl:

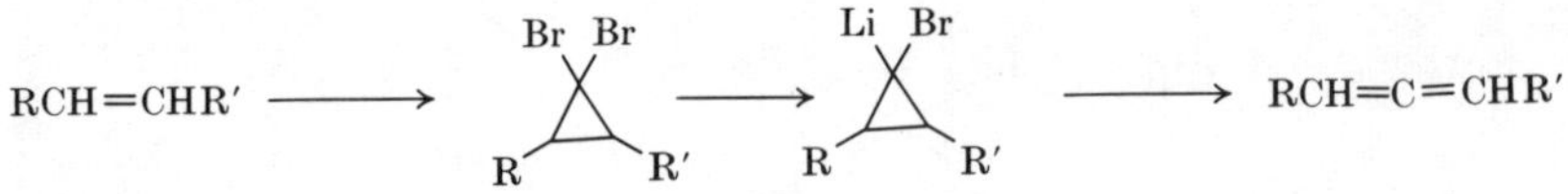

RCH=CHR' ⟶ [Br Br / R R'] ⟶ [Li Br / R R'] ⟶ RCH=C=CHR'

This reaction has been used as a partial asymmetric synthesis of cyclic allenes. Dodecene is converted into the dibromocyclopropane, which reacts with n-butyl-lithium in the presence of an optically active base [(—)-sparteine], which can co-ordinate to the lithium.[8] An optically active allene is formed:

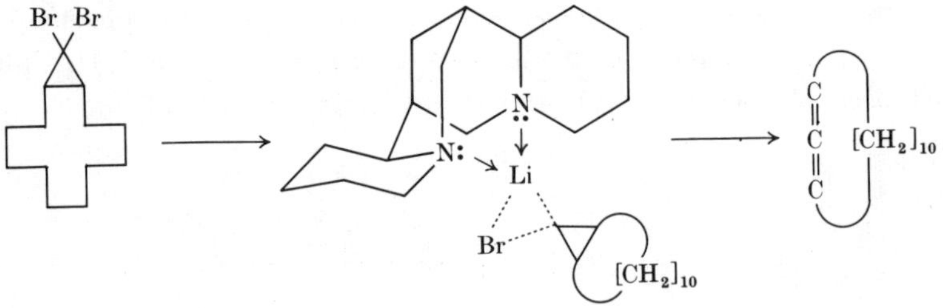

The reduction of *gem*-dihalogenocyclopropanes is also useful, as it enables monohalogenocyclopropanes and cyclopropanes to be obtained that cannot be made by the direct addition of halogeno-carbenes or of methylene to an olefin. An important example of this is the synthesis of the interesting aromatic 10 π-electron system, 1,6-methanocyclodecapentaene [**4**]. This was obtained by Birch reduction of naphthalene and addition of dichlorocarbene to the central double bond (the most nucleophilic one). The dichlorocyclo-propane was then reduced to the corresponding cyclopropane, which after bromination and dehydrobromination gave the hydrocarbon [**4**].[9]

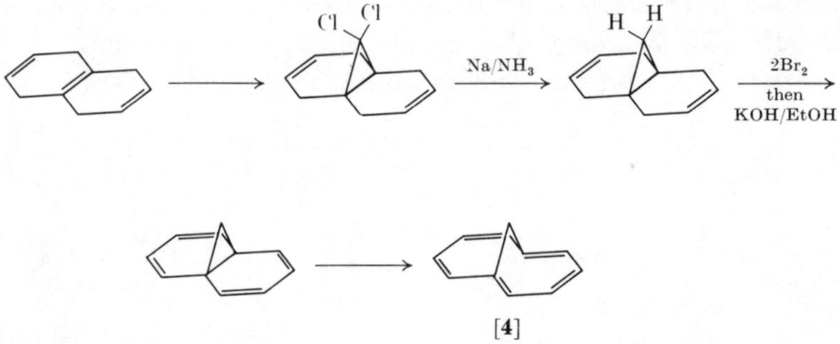

[**4**]

Dichlorocarbene has also been used as a means of introducing an angular methyl group into a steroid.[10] Addition of dibromocarbene to compound [5] (readily obtained from oestrone) gave a bis-adduct that was reduced and then ring-opened with acid. This introduced new methyl groups at C-2 and C-10 of the steroid skeleton:

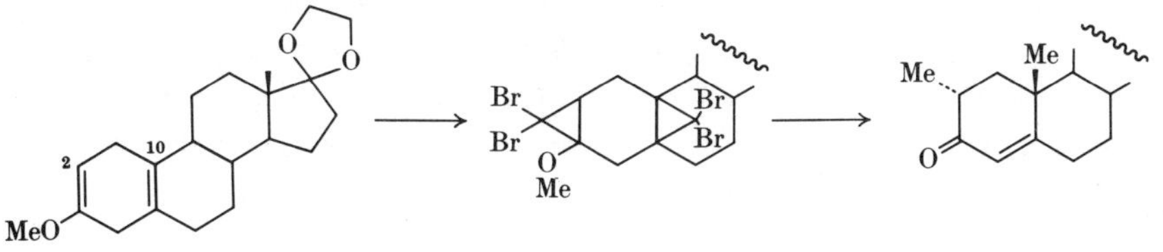

[5]

Angular methyl groups can also be introduced by means of the Simmons–Smith reaction.[11]

9–3   Diels–Alder reactions of benzyne

The applications of the Diels–Alder reaction of benzyne with dienes to the synthesis of bridged cyclic systems have already been indicated. Hydrocarbons like triptycene (from anthracene and benzyne) and benzonorbornadiene (from cyclopentadiene and benzyne) are readily made this way. The ester [6], for example, was synthesized from the corresponding cyclopentadiene ester as a first step in a possible route to the alkaloid songorine[12] [7]. More complex bridged

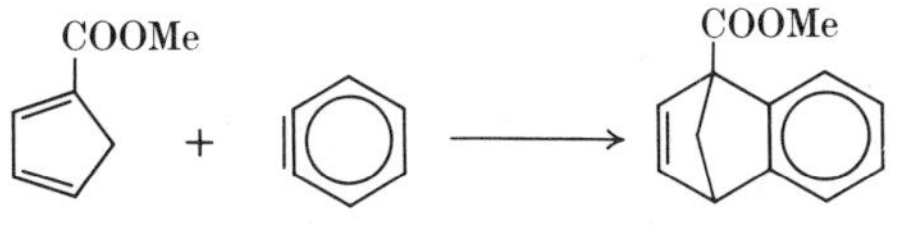

[6]

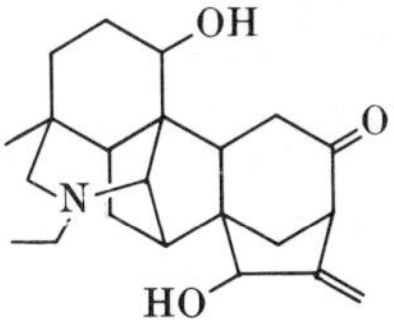

[7]

molecules are similarly available from benzyne and the appropriate diene. For example, the benzo derivative of the 'homobarrelenone' system can be synthesized in good yield from benzyne and tropone.[13] With the dimethylfulvene [8], benzyne gives the norbornadiene derivative [9][14].

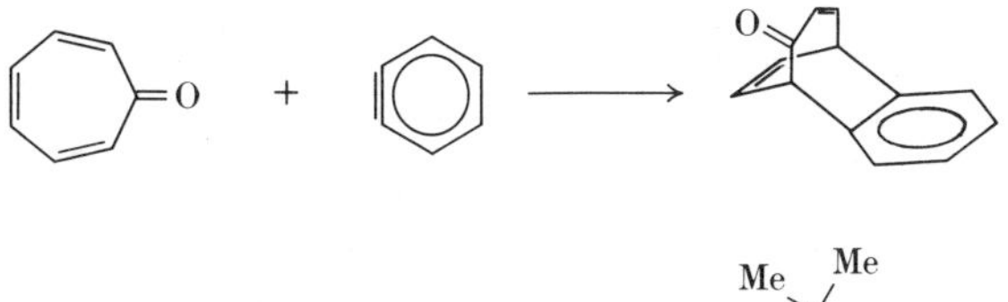

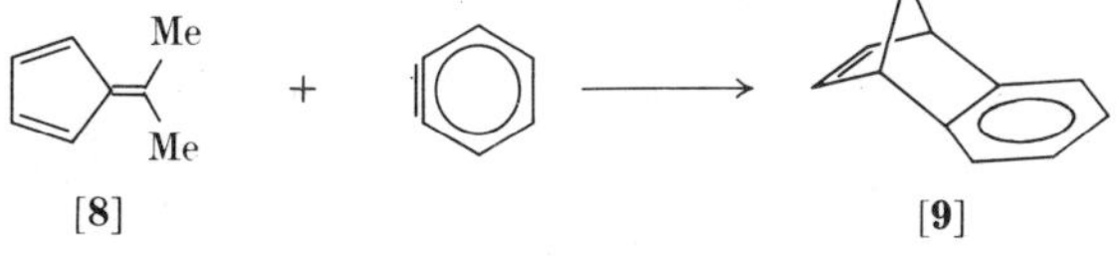

[8]   [9]

Diels–Alder reactions with benzyne can also be a good way of synthesizing more conventional ring systems. Substituted phenanthrenes can be made by the reaction between benzyne and an arylacetylene or a stilbene. With phenylacetylene the yield of phenanthrene is small, and 1,2-cycloaddition is a major competing reaction, but, with some substituted acetylenes and stilbenes, phenanthrenes are produced in good yields.[15] A general synthesis of naphthalenes is illustrated by a useful preparation of octamethylnaphthalene.[16]

Phthalazine derivatives are readily synthesized in good yields from symmetrical tetrazines, such as diphenyltetrazine [**10**], which is itself made from hydrazine and benzonitrile.[17]

9–4 Intramolecular cyclizations via aryne intermediates

The addition to benzyne of a nucleophile that is part of a side chain on the aryne, to form a new ring, was outlined in Chapter 8. It has provided a simple synthetic route to a variety of carbocyclic and heterocyclic systems fused to benzene. The reactions are usually carried out with potassamide in liquid ammonia, the side-chain nucleophile is a carbanion or nitrogen, oxygen or sulphur anion, which is usually produced *in situ* in the strongly basic medium.

The reaction provides a useful synthesis of common ring systems, as the following examples show:[18,19,20]

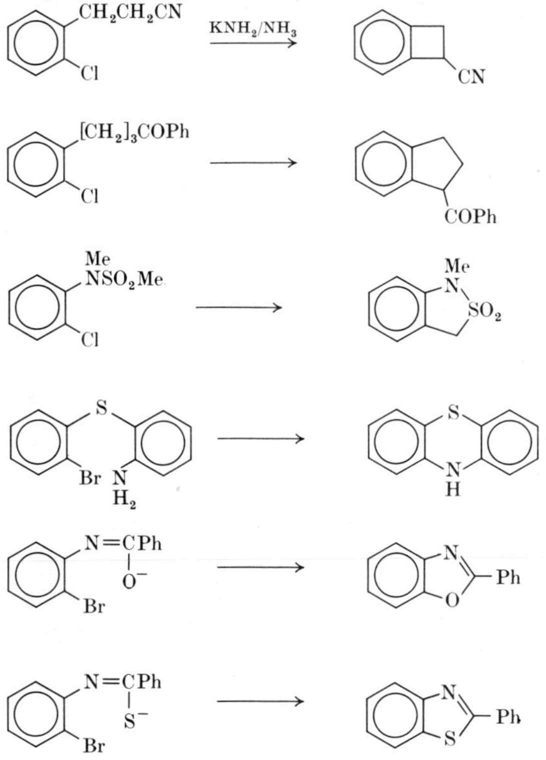

The reaction is readily adapted to more complex systems. It has been used to synthesize the tricyclic compound [11], which has a ring system related to that of lysergic acid:[21]

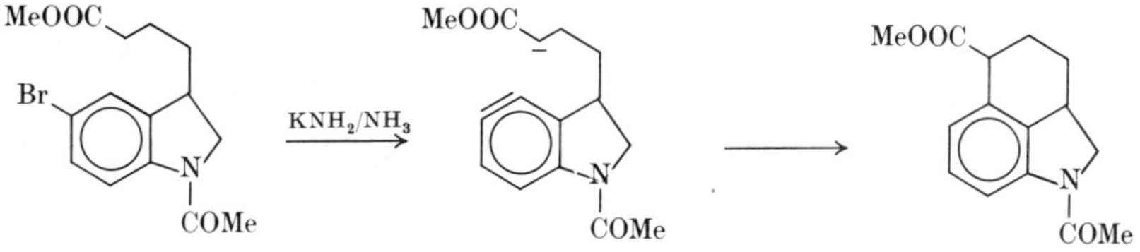

The spiro compound [12] was also prepared by this route after more conventional synthetic methods had failed:[22]

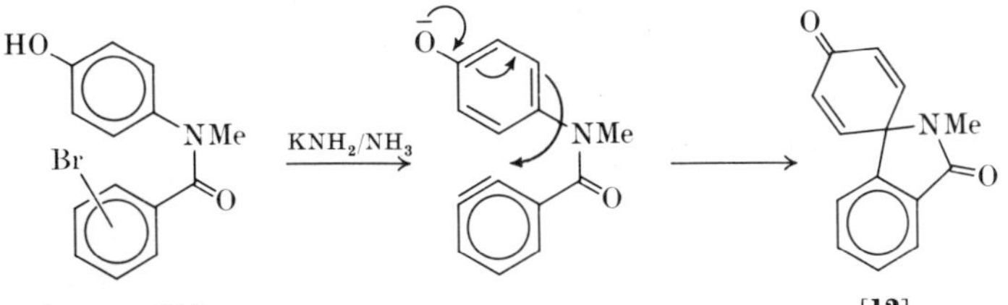

(o- or m-Br)

[12]

Novel ring systems can also be made this way: Wittig has synthesized
'azatriptycene' [**13**] by the reaction shown,[23] and Huisgen has used
halogenobenzenes with long side chains to make *meta-* and *para*-fused
heterocyclic compounds:[24]

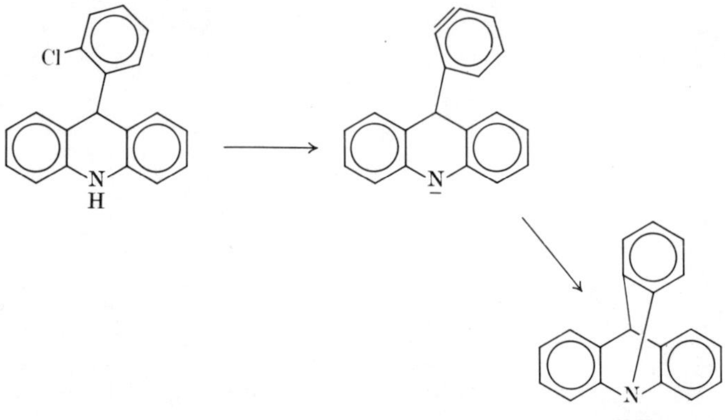

[**13**]

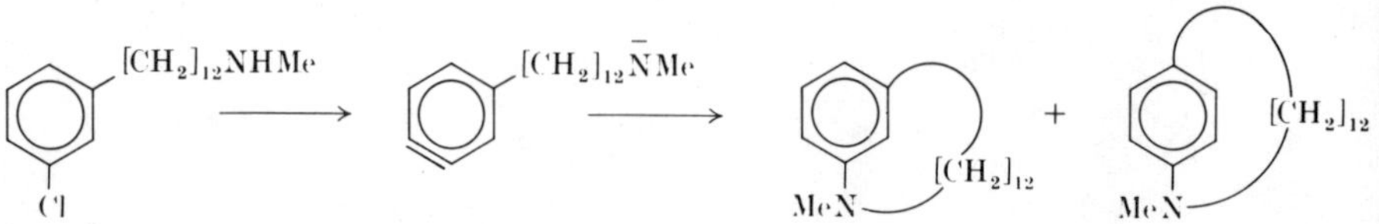

9–5   *cine*-Substitution via arynes

The *cine*-substitution that occurs in aryne reactions can be put to
good synthetic use in making isomers that are not readily available
by conventional means. For example, *o*-halogenoanisoles are readily
converted into the *m*-amines via aryne intermediates:

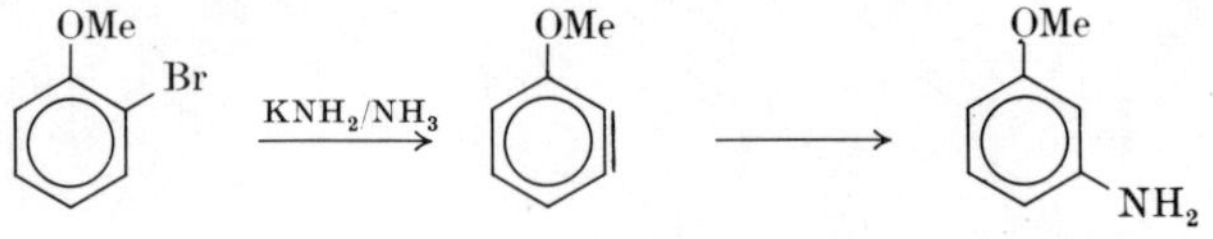

A *cine*-substitution of this sort has been used as a key step in a
synthesis of the alkaloid laureline[25] [**14**]:

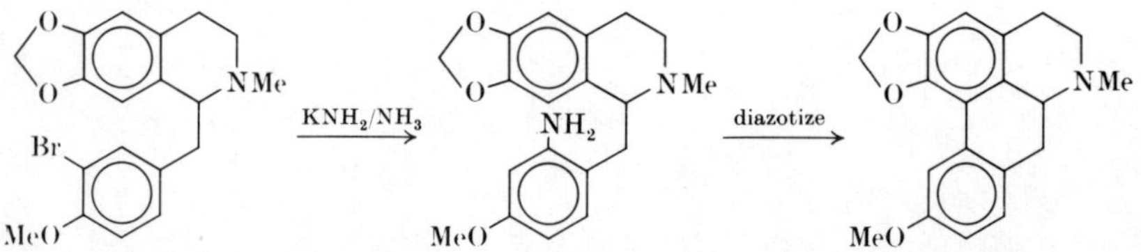

[**14**]

References

1. W. E. Parham and E. E. Schweizer, *Organic Reactions*, Vol. 13, 1963, 55, Wiley, New York.
2. P. S. Skell and S. R. Sandler, *J. Amer. Chem. Soc.*, 1958, **80**, 2024.
3. A. J. Birch, J. M. H. Graves, and F. Stansfield, *Proc. Chem. Soc.*, 1962, 282.
4. G. M. Iskander and F. Stansfield, *J. Chem. Soc.*, 1965, 1390.
5. E. E. Schweizer and W. E. Parham, *J. Amer. Chem. Soc.*, 1960, **82**, 4085.
6. W. E. Parham and R. J. Sperley, *J. Org. Chem.*, 1967, **32**, 926.
7. W. E. Parham and J. K. Rinehart, *J. Amer. Chem. Soc.*, 1967, **89**, 5668.
8. H. Nozaki, T. Aratani, and R. Noyori, *Tetrahedron Lett.*, 1968, 2087.
9. E. Vogel and H. D. Roth, *Angew. Chem. Internat. Edn.*, 1964, **3**, 228.
10. A. J. Birch, J. M. Brown, and G. S. R. Subba Rao, *J. Chem. Soc.*, 1964, 3309.
11. J. J. Sims, *J. Org. Chem.*, 1967, **32**, 1751.
12. K. Wiesner and A. Philipp, *Tetrahedron Lett.*, 1966, 1467.
13. J. Ciabattoni, J. E. Crowley, and A. S. Kende, *J. Amer. Chem. Soc.*, 1967, **89**, 2778.
14. R. Muneyuki and H. Tanida, *J. Org. Chem.*, 1966, **31**, 1988.
15. S. F. Dyke, A. R. Marshall, and J. P. Watson, *Tetrahedron*, 1966, **22**, 2515.
16. A. Oku, T. Kakihana, and H. Hart, *J. Amer. Chem. Soc.*, 1967, **89**, 4554.
17. J. Sauer and G. Heinrichs, *Tetrahedron Lett.*, 1966, 4979.
18. J. F. Bunnett and J. A. Skorcz, *J. Org. Chem.*, 1962, **27**, 3836.
19. J. F. Bunnett, T. Kato, R. R. Flynn, and J. A. Skorcz, *J. Org. Chem.*, 1963, **28**, 1.
20. J. F. Bunnett and B. F. Hrutfiord, *J. Amer. Chem. Soc.*, 1961, **83**, 1691.
21. M. Julia, F. LeGoffic, J. Igolen, and W. Baillarge, *Compt. Rend.*, 1967, **C264**, 118.
22. D. H. Hey, J. A. Leonard, and C. W. Rees, *J. Chem. Soc.*, 1963, 5266.
23. G. Wittig and G. Steinhoff, *Annalen*, 1964, **676**, 21.
24. R. Huisgen, H. König, and A. R. Lepley, *Chem. Ber.*, 1960, **93**, 1496.
25. M. S. Gibson and J. M. Walthew, *Chem. and Ind.*, 1965, 185.

General references

Carbenes

W. Kirmse, *Carbene Chemistry*, 1964, Academie Press, New York.
J. Hine, *Divalent Carbon*, 1964, Ronald Press, New York.
A. Ledwith, *The Chemistry of Carbenes*, 1964, Royal Inst. Chem. Lecture Series, No. 4.
J. I. G. Cadogan and M. J. Perkins in *The Chemistry of Alkenes*, ed. Patai, 1964, John Wiley, New York.

Nitrenes

R. A. Abramovitch and B. A. Davis, *Chem. Rev.*, 1964, **64**, 149.
L. Horner and A. Christmann, *Angew. Chem. Internat. Edn.*, 1963, **2**, 599.
W. Lwowski, *Angew. Chem. Internat. Edn.*, 1967, **6**, 897.
P. A. S. Smith, *Open-Chain Nitrogen Compounds*, Vols. 1 and 2, 1966, Benjamin, New York.

Arynes

R. W. Hoffmann, *Dehydrobenzene and Cycloalkynes*, 1967, Academic Press, New York.
Th. Kauffmann, *Angew. Chem. Internat. Edn.*, 1965, **4**, 543.

Annual reviews of new work on carbenes, nitrenes, and arynes from 1965 onwards are contained in *Organic Reaction Mechanisms*, by B. Capon, M. J. Perkins, and C. W. Rees, John Wiley, London.

Index